BSCS BIOLOGY
An Ecological Approach

EIGHTH EDITION

BSCS Green Version

A Resource Book of
Learning Skills Activities

Innovative Science Education
founded 1958

STUDENT STUDY GUIDE

BSCS
Pikes Peak Research Park
5415 Mark Dabling Blvd.
Colorado Springs, CO 80918

William J. Cairney, BSCS, Revision Coordinator
Barbara Andrews, Palmer High School, Colorado Springs, CO
Principal Author, Seventh Edition
Carol Leth Stone, Science Writer, Alameda, CA
Principal Author, Sixth Edition
Patricia Cully, Education and Editorial Consultant, Needham, MA
Janet Chatlain Girard, BSCS, Art Coordinator
Kenneth G. Rainis, Ward's Natural Science Establishment, Inc.
Linda K. Ward, BSCS, Executive Assistant

 KENDALL/HUNT PUBLISHING COMPANY

Contents

Contents

Chapter 16

16.1 Approaching Chapter 16   **125**

16.2 Writing a Short Bibliography   **126**

16.3 Vitamin C   **127**

16.4 Temperature Regulation Concept Map   **127**

16.5 Reviewing   **128**

Chapter 17

17.1 Skimming for Meaning   **131**

17.2 Making a Flow Chart   **132**

17.3 A Pattern of Behavior   **133**

17.4 The Control of Blood Sugar Level   **135**

17.5 Cocaine   **136**

Chapter 18

18.1 Yarrow Growth   **137**

18.2 Transport Analogies   **138**

18.3 Sales Resistance   **139**

18.4 Plant Growth   **140**

18.5 Why Do Yarrow Plants Vary?   **141**

Your textbook presents the ideas and themes of biology, many specific details, and a large new vocabulary. At times, this new information may seem overwhelming. You can, however, take charge of your learning and become successful in biology. This guide is designed to assist in doing so.

This study guide will help you learn to use science skills and knowledge to solve biological problems. By doing such things as hypothesizing, using data, and making inferences, you will better understand how and why these skills are necessary for science. You will learn to apply your new knowledge about biology to the world around you. Memorizing the words or ideas of biology is useless unless you can apply biology to your own life. You also will learn how to communicate with others about biology in various ways.

We hope the study guide will help you master other learning, as well as biology. We want to know if this guide helps you, and if you have suggestions for changes for the next edition. Please send your comments to BSCS, Pikes Peak Research Park, 5415 Mark Dabling Blvd., Colorado Springs, Colorado 80918-3842, Attn: GV8.

Predicting Your Future

Much of the information you will study in biology is related to your own life and to your immediate surroundings. It also is related to what your life may be like in the future.

What will your future be? You can use your knowledge of the present to make predictions about the future. Be creative with your predictions, but keep them realistic. Based on what you know and what you can guess, write your answers to the following questions.

Twenty years from now, where do you think you will live?

Will you live in a city or a rural area?

Will you be married?

How many children will you have?

How will you care for your children?

Will your parents live with you?

What types of foods will you eat? Where will they come from? Will you grow any of your own food?

What sort of work will you do? Where will you do it?

What type of exercise will you get? Where will you go for it?

How healthy do you expect to be?

What will your surroundings be like? Will they be crowded and polluted, or spacious and clean?

Can you make any other predictions about your future?

Activity 1.1

Getting Ready to Read

If you read a book written for a young child, you can finish it quickly and no words or ideas will seem strange to you. Reading a book written for high school students is more difficult. The authors may use some words you do not understand, or refer to ideas you have forgotten since you first learned them. If you try to read the book without enough background information, you will waste time and you may misunderstand what you are reading.

Your teacher may spend some class time reviewing words or ideas that will help the entire class. It is up to you, however, to make sure that you are ready to read the chapter and understand it.

To get ready, skim the chapter. Do the authors refer to some things without explaining them, assuming you already understand them? In the space provided, list a few things you should look up or learn before reading Chapter 1 thoroughly.

Activity 1.2

Taking Helpful Notes

What you learn in biology may come from several sources, such as laboratory work, the textbook, and outside reading. One important source of information is your teacher. This activity will help you to write down information your teacher gives you and to use it later.

The following example is taken from one biology student's class notes. Notice what the student did while taking the notes—underlining and sketching something the teacher drew on the chalkboard.

Oct. 1

<u>Food chains + webs</u>

* Food chain - series of steps in passage
of food from one organism
to another. Always a plant,
may have several animals.

arrow - eaten by or gives   How many?
food to

plant → animal → animal → decomposers (nots)
grass → rabbit → coyote → maggots <u>yuck!</u>

* Interrelationships - ways orgs affect
each other. Can link chains
to make web.

* Food web

grass → rabbit → coyote → decomposers
      → ewe → lamb   ?

play tryouts
Oct 15!

Quiz on
Monday !

Why do you think the student added stars to some terms?

After class, the student and some friends looked over their notes together and helped each other understand the material. They looked up some unfamiliar words in the textbook and a dictionary. The student then typed the notes and put them in a permanent biology notebook. The notes about the quiz and the play tryouts went into a reminder book. The typed notes looked like this:

```
Oct. 1



                         Food chains and webs



*Interrelationships—ways organisms affect each other. Includes eating

and other relationships.


*Food chain--series of steps in the passage of food from one organism to

another. Always includes one plant; may include one animal or several, as

well as decomposers.


*Decomposer--anything that decomposes (rots) tissue, such as bacteria or

mold.


General food chain:

plant  -> animal  -> animal  -> decomposer

Examples:

grass  -> rabbit  -> coyote  -> maggots

grass  -> ewe  -> lamb  -> coyote

(Arrow = eaten by, or gives milk or other food to)


*Food web--two or more food chains with some shared organisms, such as:


grass  -> rabbit  -> coyote  -> maggots
      \                              /
       \-> ewe  -> lamb  -/
```

Reviewing and copying your notes may appear to be a lot of work, but most students who do it find they save more time later in studying for tests. Try this method on your notes from the next biology class.

Activity 1.3

Seeing with a Purpose

When you have some understanding of how organisms grow and behave, it is easier to orient yourself in your surroundings. Such orientation was very useful to Native Americans, who had no compasses. An 18th-century priest, Pere Lafitau, spent five years with the Iroquois tribe and observed they "pay great heed to their 'star' compass [the Pole Star] in the woods and in the vast prairies. . . . But when the sun or stars are not visible they have a natural compass in the trees of the forest from which they know the north by almost infallible signs.

"The first was that of their tips, which always lean toward the South, to which they are attracted by the sun. The second is that of their bark, which is more dull and dark on the north side [where there is less evaporation, and the moist bark looks darker]. If they wish to be sure they only have to give the tree a few cuts with their axe; the various tree rings which are formed in the trunk are thicker on the north side." (From Gatty, Harold, 1958, *Nature is Your Guide.* New York: Penguin Books.)

Using Iroquois methods, label the north side of each of the following pictures:

These methods are not always reliable. What might cause a tree to grow toward the north, or to be moister on the south side?

Activity 1.4

Constructing a Concept Map

Chapter 1 discusses the relationships between living organisms as a web of life. The web illustrates the interdependence of all organisms that share an environment. The food web in Figure 1.7 of your textbook has many strands that connect the various organisms within a community. These strands, or pathways, indicate the relationship between the organisms at each end of the strand.

A **concept map** is similar to a food web. It illustrates the relationships among ideas. These ideas, or concepts, center around a main concept. For example, the main concept of Figure 1.7 in your textbook might be *food web*. Each organism then becomes an idea in the concept map. The strands that connect these ideas are labeled with linking words or phrases that describe the relationship between the two concepts. The fox and the rabbit are examples of ideas on the concept map and the relationship between these two ideas could be described as predator/prey—the fox eats the rabbit. *Eat* is written on the strand that connects foxes and rabbits. Write in the terms that describe the animal relationships in the following concept map.

foxes

birds   rabbits

grasshoppers   spider

grass   raspberry
bushes

Each concept map may be slightly different. There are no right or wrong answers if you follow a few guidelines:

1. A concept map usually stems from one main idea.
2. The main idea branches into related general concepts.
3. These general concepts are subdivided into specific concepts that are expanded through examples.
4. Each concept, usually a noun, should represent a single idea and appear only once in the map.
5. The relationship between concepts is shown by a linking word, usually a verb, adverb, preposition, or verb phrase, and all concepts should be linked.
6. Cross-linkages are used to connect concepts in different areas of the map. The more cross-linkages, the better, but crossed lines should be avoided.
7. Any two concepts and their linking word(s) should form a complete thought.

The concept map you labeled does not follow all of these guidelines. What changes should be made to better illustrate the ideas stemming from the main concept *food web?* Could some other ideas be added to help subdivide the concepts? Reconstruct the concept map so that it follows the guidelines and gives a clear illustration of the relationships in this food web.

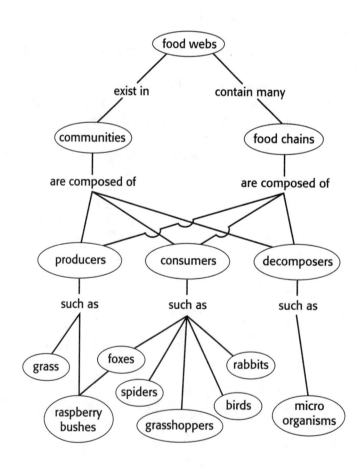

Activity 1.5

Disappearing Marshlands

San Francisco Bay was surrounded by marshes when the Spanish explorer Gaspar de Portolo arrived in 1769. He found the Ohlone Indians hunting deer, rabbits, and other game in the tall marsh grasses.

Through the years, many of the marshes have been drained for various human purposes. Some people regard the marshes as wet, smelly places that lower property values. Many animals and plants, however, live only in the marshes, and many others depend on organisms that live there.

About three dozen types of birds live year-round in the marshes including ducks, grebes, coots, killdeer, avocets, stilts, and clapper rails. They feed on small animals that live in the mud—worms, snails, and shellfish (clams, mussels, oysters, and shrimp). These small animals, in turn, eat very small marsh producers such as diatoms and algae. They also eat decaying animal and plant matter called detritus.

Much detritus comes from plants that can tolerate high concentrations of salt, such as cordgrass, salt grass, marsh rosemary, alkali heath, pickleweed, and Australian saltbush. In addition to providing materials as detritus, pickleweed acts as host to a parasitic plant called salt marsh dodder, and the leaves of the Australian saltbush are the food of the caterpillars of pygmy blue butterflies, the smallest butterflies in North America.

Land birds such as sparrows, meadowlarks, and blackbirds come to the marsh to feed on insects, and they may become food for larger birds including kites, short-eared owls, and marsh hawks. In the mud along the water's edge, crabs and shrimp provide food for migratory birds such as sandpipers, various types of ducks, great blue herons, and great egrets. The herons and egrets also feed on mice. One type of mouse, the red-bellied salt marsh harvest mouse, is an endangered species that lives nowhere else. The mouse is food for hawks, owls, herons, and gulls.

Humans fish for striped bass, surfperch, Pacific herring, sturgeon, and flounder in the bay, as well as harvesting its shellfish. Fish feed on plants, snails, and small shellfish.

In the space provided below, draw a food web based on the information presented above.

The marshes now are restricted to about 133 km² of open space. Many people using the area would like to drain some of the marshy areas that remain, making them suitable for such uses as airport runways. What might be the effects of such drainage?

Activity 2.1

Writing about Populations

Get out a pen or pencil and several sheets of scratch paper. Glance through Chapter 2 in your textbook for a minute or two.

Close the book and, for the next five minutes, write down what you know about populations. Include topics from the textbook and topics you know about from other sources. Do not worry about spelling or grammar; just write freely, getting your ideas on paper. Do not worry about whether your ideas are correct. Write any questions you have about populations on a separate sheet of paper or underline them to make them stand out from the rest of what you have written.

When the time is up, go back and organize what you have written. Write at least three paragraphs, corrected for spelling and grammar, to show your teacher. Write your questions in the space provided below. Your teacher may ask to see the questions now, or you may wish to raise them in class discussions.

Activity 2.2

Planning How to Study Biology

By now, you have some idea of what to expect from your biology course. How can you take charge of your work so that you will learn as much as you can?

You need to know what the assignments are and when they are due. Your teacher will tell you about the reading assignments and the tests, as well as any other work that is expected of you. From that point, it is up to you.

One thing that will help you in your studies is planning ahead. You need to estimate the time needed for reading, doing the assignments, doing problems, studying for tests, and so on. Based on what you have experienced so far this school year, fill out this tentative weekly schedule:

Activity	Amount of time	Day and time
Reading biology assignments		
Reviewing biology assignments		
Studying for biology tests		
Other (extra projects, etc.)		

The amount of time and how you schedule it will vary from person to person. For example, if you are a slow reader you should plan to spend more time on the assignments than would a faster reader. If you can do math problems more easily, it will help in doing some of the work that lies ahead. If you have a job or extracurricular activities after school, you will have to juggle your time accordingly. Set up a schedule you think will work for you. For two or three weeks, keep track of the time you actually spend and revise the schedule if needed.

In addition to planning your time, you need to plan your approach. In the space provided, write a short paragraph about how you plan to read a chapter to locate and remember the main ideas in it.

Activity 2.3

Pig Populations

Two groups of pigs were kept in adjacent pens. The fences around the pens were high and strong, but the hedge separating the two groups of pigs was not strong enough. Sometimes pigs were able to get through the hedge.

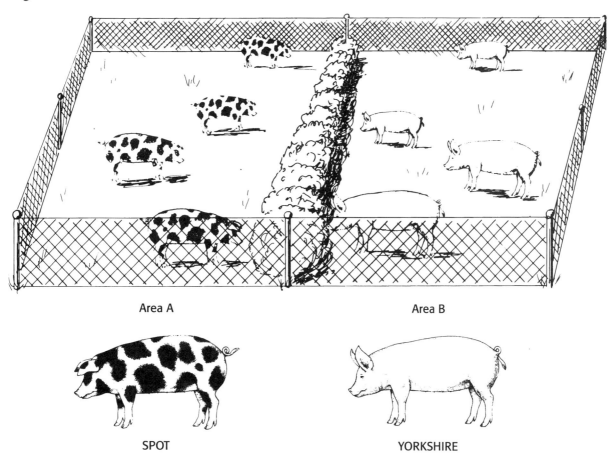

Area A   Area B

SPOT   YORKSHIRE

At the beginning of the year, all the pigs in Area A were of the Spot breed, as shown on the left. All the pigs in Area B were of the Yorkshire breed, shown on the right. Events during Year 1 are summarized in the following table:

	Area A	Area B
Year I starting population size	6	9
Number of births	18	21
Number of deaths	6	9
Number of emigrations	9	11
Number of immigrations	11	9

At the end of Year 1, how many pigs were in each area?

	Area A	Area B
Spot		
Yorkshire		
Total		

Events during Year 2 are shown in the following table:

	Area A	Area B
Year 2 starting population size		
Number of births	55	47
Number of deaths	20	19
Number of emigrations	25	27
Number of immigrations	27	25

At the beginning of Year 3, none of the pigs looked like those shown above. Most of them looked more like these pigs:

Events during Year 3 are shown in the following table.

	Area A	Area B
Year 3 starting population size		
Number of births	285	225
Number of deaths	70	60
Number of emigrations	100	50
Number of immigrations	50	100

How many pigs were in Area A at the end of Year 3?

In Area B?

What rate primarily was responsible for the changes in population sizes during Year 3?

Try to account for this.

Activity 2.4

Using Tables of Data

Most science articles, and many newspaper articles, contain tables of data, or information. At first, these tables may appear to be confusing columns of numbers. If you learn to use such tables, however, you can acquire and use a large amount of information quickly. If the information were written in paragraph form, it would take up a great deal of space and could not be used as easily.

The simple table below shows the results of growing maple seedlings in different solutions. The left column shows the original dry weight of each seedling. The other columns show what each seedling weighed after two months of growth on sphagnum moss, water, and other nutrients.

Dry weight (grams) after growth in solutions containing:

Original dry weight in gm	Water only	Phosphorus (P) potassium (K); no nitrogen (N)	N and K; no P	N and P; no K	N, P, and K
0.038	0.077	0.071	0.077	0.490	0.423

Which element(s) contributed greatly to seedling growth?

Which element(s) is/are not essential to seedling growth?

Use the following paragraph to make a data table of your own.

A research analyst at the Population Council conducted a study of death rates due to heart attacks in women ages 40 to 44. He found that for American and British women in that age range, the annual death rate for women who neither smoked nor used oral contraceptives was 7.4 per 100 000. On the other hand, women ages 40 to 44 who did not smoke, but who did use "the pill" had a death rate of 10.7 per 100 000. For women who smoked but did not use the pill, the rate was 15.9. The death rate was 62 per 100 000 in women who both smoked and used the pill.

Activity 2.5

Population Profiles

Some observations are direct. You can look at a pig, for example, and see whether it is white or black, spotted or unspotted. Other people probably will make the same observation.

Other observations are indirect and may leave room for disagreement. Such indirect observations are called inferences. By observing something, you infer that something else is true. If you observed that the lock on your front door was broken, you might infer that a burglar had entered the house.

Biologists and other researchers often use population data to infer other things about a population. Here are some data about three human populations, arranged by age group only:

Age Group	Population 1	Population 2	Population 3
0–10	30%	20%	5%
10–20	25%	15%	5%
20–45	30%	35%	35%
45–65	10%	20%	35%
65+	5%	10%	20%
Total	100%	100%	100%

Which population is most likely to grow rapidly in size over the next 20 years, and why?

Which population probably has the greatest number of men between the ages of 45 and 65?

Make at least one inference from the table. See if others in the class agree with your inference. If you disagree, how can you find out who is right?

Chapter 3

Activity 3.1

Getting an Overview

How do you begin reading an assignment in your textbook or other source of information? If you just begin reading and plod through the assignment to the end, you may lose track of what is happening somewhere in the middle. You will learn faster and learn more if you skim the assignment first to get a rough idea of the topics covered. By looking at each page for just a few moments, you can get a broad overview.

You also probably will forget your overview quickly unless you write down what you remember. It is not necessary to write the information down in complete sentences or to worry about spelling. The important idea is to write down everything you can immediately recall. Here is an example of what one person wrote down after skimming Sections 2.5 through 2.11 in the textbook.

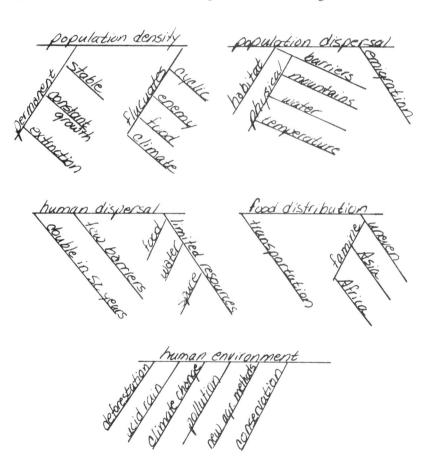

Notice that connecting lines were used to link related words. The student recalled that population density fluctuates through time because of various factors. Illustrations could have been sketched or colored pencils used to emphasize certain ideas or links. There is no one way to record what you recall from skimming information. You may use whatever method seems appropriate at the time.

 Try skimming Chapter 3 in your textbook. Look at each page for several seconds. Remember that you are going to record what you can recall. There are no right or wrong answers. If you make mistakes, revise your ideas as you read the chapter more thoroughly. Write your notes in the space provided.

Activity 3.2

Writing an Outline

Outlines are useful for two main purposes: summarizing what someone else has written, and planning what you are going to write.

Outlining what others have written is especially useful if you wish to criticize their writing. Scientists often exchange rough drafts of their papers before sending them to publishers. If you are asked to criticize someone's rough draft, you can help her or him organize the paper by outlining it.

Outlining a paper you are planning is also useful for organization and self-criticism. Using an outline to organize your thoughts before you begin writing helps ensure you have included all the major points you wish to make. You also can check to see whether you have included topics that are unimportant or inappropriate for this particular paper.

The following outline is based on Chapter 3 in your textbook:

Communities and Ecosystems

A. Introduction

B. Life in a Community

    3.1 Many interactions are indirect

    3.2 Many populations interact in the Florida River community

    3.3 A niche is the role of an organism

    3.4 Organisms may benefit or harm one another

C. Ecosystem Structure

    3.5 Boundaries of ecosystems overlap and change

    3.6 Most communities have more producers than consumers

D. Ecosystem Stability and Human Influences

    3.7 Humans lower the stability of ecosystems

    3.8 Humans threaten the diversity of organisms

    3.9 Humans also can help conserve species

Do any of the numbered headings seem out of place?

Look at the paragraphs in that section. Do you understand its placement now?

If you were writing an outline for a textbook and came across such a problem, how would you solve it?

Why do you think the section headings are written in the form of complete sentences?

Activity 3.3

Finding a Niche

The plant *Dioclea* grows in Kentucky. It is a legume, or pod-bearing plant. Unlike peas or beans, however, it is poisonous to most, but not all, insects. A certain beetle, *Caryedes,* eats nothing else.

   The beetle's life cycle is completely dependent on *Dioclea.* The immature beetles, or larvae, grow inside the plant's seeds. When they become adults, the beetles emerge from the seeds and feed on the plant's pollen. They lay eggs on the seed pods. The eggs hatch into larvae, which burrow into the seeds, and the cycle repeats.

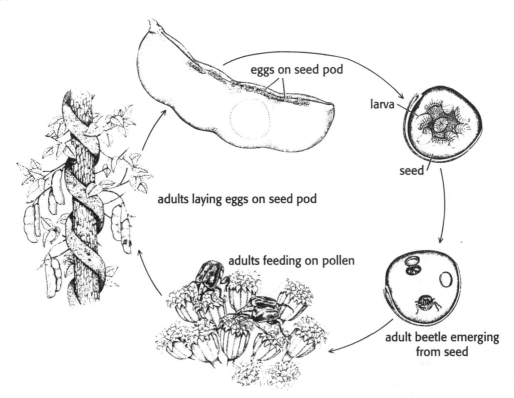

eggs on seed pod

larva

seed

adults laying eggs on seed pod

adults feeding on pollen

adult beetle emerging
from seed

Another insect in Kentucky is the tobacco hornworm. It cannot eat *Dioclea,* but it feeds on tomatoes, tobacco, and other plants.

Draw the food chains described for *Caryedes* and the tobacco hornworm.

Describe the niches occupied by *Dioclea* and *Caryedes.*

What advantages and disadvantages are there to this arrangement for *Dioclea?*

What advantages and disadvantages are there to this arrangement for *Caryedes?*

Activity 3.4

A Problem of Immunity

A scientist read a journal article about a village near the east coast of Africa. A parasite that lives symbiotically with 70% of Africans is never found in the villagers. The villagers are very fond of drinking a root beer they make from a local plant that grows only there.

   The scientists made a tentative inference about the villager's apparent immunity to the parasite. What was it?

Another scientist read another article about the parasite. The article said nothing about the root beer, but with the article was a rough map of the village and the surrounding area.

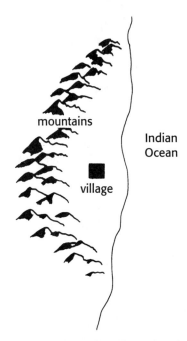

That scientist also made a tentative inference about the villager's immunity to the parasite. What was it?

The two scientists met each other at a conference and they discussed their inferences about the village. They decided to travel there together to discover the real cause of the villager's immunity. What preliminary study might they do in Africa?

Chapter 4

Activity 4. 1

Using Models

Part A: Bohr Model of the Atom

A Bohr model helps to visualize and understand the atom. The nucleus of the atom contains the positively charged protons (P) and the neutral neutrons (N). The protons and neutrons together make up the atomic mass of an atom. The negatively charged electrons (E) circle around the nucleus in orbitals, energy levels, shells, or electron clouds. The number of electrons and the number of protons in an atom of a given element determine its atomic number.

The name of each element, its symbol, atomic number, and atomic mass are found in the periodic table of elements. With this information, it is possible to construct a model or diagram that illustrates the atomic structure of an element.

Hydrogen is the simplest atom. It has an atomic number of one and an atomic mass of one—it has one proton and one electron. It does not have a neutron. The Bohr model for a hydrogen atom has one proton in the nucleus and one electron in the orbital closest to the nucleus. The following figure shows the Bohr model for hydrogen.

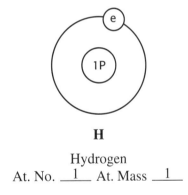

H

Hydrogen
At. No. __1__ At. Mass __1__

Helium is the next element on the periodic table of elements. Helium has an atomic number of two and an atomic mass of four. Based on this information, how many protons would you expect in the nucleus?

Would any neutrons be present?

How many electrons would be present?

Complete the model for helium.

He

Helium

At. No. _____ At. Mass _____

As the atoms become larger for other elements, the electrons begin to fill the orbitals. The inner-most orbital is smallest and can hold only two electrons. When the orbital is filled, it is in a stable state. The next two orbitals are larger and can hold up to eight electrons each in a filled, stable state. The inner orbitals must be filled before an electron can occupy the next orbital out from the nucleus.

      Hydrogen, carbon, nitrogen, and oxygen are the most common elements in biological molecules. Draw the Bohr models for these elements. Refer to the periodic table of elements in Appendix 3 of your textbook.

C	**N**	**O**
Carbon	Nitrogen	Oxygen
At. No. __6__	At. No. _____	At. No. __8__
At. Mass __12__	At. Mass __14__	At. Mass _____

Several other elements are important in biological molecules. Use the periodic table to help you draw the Bohr models of these other elements.

Na

Sodium
At. No. _____
At. Mass _____

P

Phosphorus
At. No. _____
At. Mass _____

S

Sulfur
At. No. _____
At. Mass _____

Cl

Chlorine
At. No. _____
At. Mass _____

Part B: Models of Atoms Combining to Make Molecules

Part A helps illustrate the structure of atoms. The electrons inhabit the outer part of the atom and, therefore, are the part of the atom that reacts or combines with other atoms to make molecules. The outer orbital of an atom tends to reach a stable state by losing, gaining, or sharing electrons with other atoms.

The formation of charged atoms is called **ionization.** In the space provided, complete the Bohr models for the outer orbitals of sodium and chlorine atoms, and give the net charge of each ion after the movement of the electron. Draw an arrow indicating the movement of an electron from one atom to the other that results in stable outer orbitals for both atoms. This force creates an **ionic bond** between the atoms.

Na	**Cl**
At. No. _____	At. No. _____
Net charge __+__	Net charge __−__

Would these oppositely charged ions be attracted to or repelled by each other?

Which atom is most likely to give up an electron?

Is the other atom likely to except another electron?

Do the positive charges in the nucleus and negative charges outside the nucleus in each atom balance after the movement of the electron?

Complete the Bohr model of the outer orbital of hydrogen and chlorine. Draw an arrow indicating the ionic bond formed in the molecule of hydrochloric acid. Give the net charge of the ions formed after the movement of the electron.

H   **Cl**

At. No. _____   At. No. _____

Net charge __+__   Net charge __−__

Not all electron movement results in ionization. In the combination of oxygen and hydrogen atoms that form the water molecule, how many electrons are need to stabilize each hydrogen atom?

How many electrons are needed to stabilize the oxygen atom?

Can the total of eight electrons be redistributed among the atoms to satisfy each atom's requirement to stabilize, and is there a net charge on the atoms after redistribution?

Draw arrows indicating the redistribution of electrons in a water molecule.

Complete the Bohr model of the outer orbital for each atom indicated below. Use double headed arrows to indicate the sharing of electron pairs, a **covalent bond,** to give each atom of the molecule filled, stable outer orbitals. More than one pair of electrons may be shared between atoms.

Ammonia (NH$_3$)   **Carbon Dioxide (CO$_2$)**

The molecules of acids and bases may react with other atoms by breaking an ionic bond. What ions are produced by the ionization of hydrogen and chlorine atoms that make up hydrochloric acid?

The hydrogen ion is made of what particle?

This proton is the hydronium ion common to all acids. Not all bonds in acids and bases are ionic bonds. Covalent bonds tend to stay intact while ionic bonds may break in solution with water. Water also can be ionized. What other ion would be formed when the hydronium ion separates from the water molecule?

This hydroxide ion is common to all bases. What ions would be formed when sodium hydroxide is dissolved in water?

Complete the Bohr models for the ionization of water and sodium hydroxide.

water

H   Ion _____   Ion _____
     Net ch. _+_   Net ch. _−_

sodium hydroxide

Ion _____   Ion _____
Net ch. _+_   Net ch. _−_

Make models of some other acids and bases you have studied in class and indicate the bonds and ions that they form in solution.

Activity 4.2

The Vernacular of Science

You probably speak only one or two languages, but may use a different vernacular, or way of speaking, with your friends, teachers, and family. Science has a vernacular of its own. You will be exposed to as many new words and concepts in biology this year as you would in the first year of a foreign language. Fortunately, there are some techniques you can use to learn these words quickly, without just memorizing each new term.

1. Many prefixes and suffixes are added to root words to show variations in meaning. If you know what a prefix or suffix means, you can recognize it in combination with any root word. For example, the prefix bio- means "life" or "living," and the suffix *-ology* means "the study of." List some other words that begin with *bio-,* and write their definitions.

List some words that end with *-ology,* and give their definitions.

Look through the glossary and index at the end of your textbook. Choose some of the prefixes and suffixes that appear more than once and see if you can determine what they mean. List them below and check your answers in a dictionary.

Prefix	Meaning

Suffix	Meaning

If you found the following made-up words in an article, how would you define them?

Erythrophyll _____

Anthrovore _____

Neozoic _____

Oxymolecular _____

Make up some words yourself and see if other students can translate them.

2. Knowing the original meaning of a scientific word can help you remember its present meaning. Use an unabridged dictionary to look up the derivations of the following terms and write them down in the space provided.

Epidemic _____

Vector _____

Niche _____

Opiate _____

3. Try to study in a place where you can use the dictionary. If you look up an unfamiliar word immediately, you will better understand what you are reading. If no dictionary is available, write the word down and look it up later.

Activity 4.3

Giving Instructions

Assume you are tutoring a student in biology. Draw a chart for your student to use in studying biological molecules. The rows across should be the four major types of biological molecules. The columns down should be characteristics, functions, sources, and so on, that will help your student understand and distinguish these molecules from one another. Give instructions for filling out the chart, but do not fill it in. You may wish to fill it in yourself as you study and review Chapter 4.

Activity 4.4

Constructing a Concept Map

Concept maps help create a mental image of how ideas relate to each other. General concepts and ideas are used as building blocks for the map. These concepts and ideas are connected by phrases that describe the relationship between the connected ideas. For example:

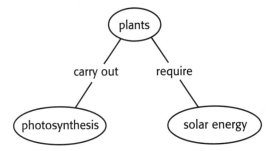

Here, the main concept of plants is related to the concepts of energy and photosynthesis. The connecting phrases indicate the relationship between the concepts.

Chapter 4 discusses several ways in which energy fits into the web of life. Construct a concept map using the study chart you made in Activity 4.3. The foundation is provided below. Relate the biological molecules of carbohydrates, proteins, and lipids to the concept of energy. Add additional concepts and ideas as you need them.

Activity 4.5

Variables

A variable is anything that changes. Scientists study some things that vary naturally, such as population size or the pH of water. When they plan experiments, they control variables and determine how changing each variable changes the outcome of the experiment.

   Many natural variables are found in Chapter 4 of your textbook. For example, every molecule varies in the atoms that make it up. For each of the following items, what may vary?

Adenosine phosphates _____

Carbohydrates _____

Proteins _____

Enzyme activity _____

   The variable that changes first in a system is called the **independent** variable and it usually is the variable controlled by the experimenter. If something else changes because it depends on the independent variable, it is called a **dependent** variable. During a study of population growth, for example, the time period of the study is the independent variable controlled by the experimenter. The change in size of the population during the time period under study is the dependent variable.

   In the following descriptions of experiments, determine the independent and dependent variables.

1. A scientist is interested in the rate of reaction of a new enzyme. He plans to consider temperature, pH, enzyme concentration, several possible substrates, and different substrate concentrations to determine the optimum conditions of the enzyme.

   Independent variable(s):

   Dependent variable(s):

2. A new bacterium has been discovered and the scientist wants to learn more about its nutritional requirements. He attempts to grow the bacteria on agar plates containing different nutrient sources: glucose, starch, blood, lipids, and minimal nutrients. He incubates a set of these plates inoculated with the bacteria at room temperature, with one plate placed in the dark and the other plate placed under a light source. Two additional sets are incubated at 37° C in the dark and in the light.

Independent variable(s):

Dependent variable(s):

In testing new drugs for effectiveness, chemists for drug companies may vary portions of drug molecules. For example, here is the molecular formula for penicillin:

The portion labeled "R" can be modified to form other penicillins, such as staphcillin or ampicillin. Write a paragraph describing how new penicillins might be tested for effectiveness against a new strain of bacteria. Correctly use the terms "independent variable" and "dependent variable" in the paragraph.

Chapter 5

Activity 5.1

The Main Idea

In well-written science material, each paragraph has a main idea. Usually that idea is stated in one sentence, called the topic sentence. The topic sentence often is the first sentence in the paragraph. The rest of the paragraph develops the main idea further by providing examples making comparisons, giving details, or adding information to the topic sentence in other ways.

Read the following paragraph, written in 1674 by van Lecuwenhoek:

About two hours distant from this Town there lies an inland lake, called the Berkelse Mere, whose bottom in many places is very marshy, or boggy. Its water is in winter very clear, but at the beginning or in the middle of summer it becomes whitish, and there are little green clouds floating through it; which, according to the saying of the country folk dwelling thereabout, is caused by the dew, which happens to fall about that time, and which they call honey-dew. This water is abounding in fish, which is very good and savory. Passing just lately over this lake, at a time when the wind blew pretty hard, and seeing the water above described, I took up a little of it in a glass phial; and examining this water next day, I found floating therein divers earthy particles, and some green streaks, spirally wound serpentwise, and orderly arranged, after the manner of copper or tin worms, which distillers use to cool their liquors as they distil over. The whole circumference of each of these streaks was about the thickness of a hair on one's head. Other particles had but the beginning of the foresaid streak; but all consisted of very small green globules joined together: and there were very many small green globules as well. Among these there were, besides very many little animalcules, whereof some were roundish, while others, a bit bigger, consisted of an oval. On these last I saw two little legs near the head, and two little fins at the hindmost end of the body. Others were somewhat longer than an oval, and these were very slow amoving, and few in number. These animalcules had divers colours, some being whitish and transparent; others with green and very glittering little scales; others again were green in the middle, and before and behind white; others were yet ashen grey. And the motion of most of these animalcules in the water was so swift, and so various, upwards, downwards, and round about, that 'twas wonderful to see: and I judge that some these little creatures were above a thousand times smaller than the smallest ones I have ever yet seen, upon the rind of cheese, in wheaten flour, mould, and the like. (From Dobell, Clifford. *Antony van Leeawenhoek* and *His 'Little Animals.' 1958, New York: Russell and Russell.*)

What is the main idea in the paragraph?

If the paragraph has a topic sentence in it, underline it. How is the paragraph developed?

Activity 5.2

Identifying a Hypothesis

A hypothesis sometimes is called an educated guess. All of us make informal hypotheses based on our past experiences and knowledge. For example, if you observed a great many sparrows and each of them had white crossbars on its wings, you might hypothesize that all sparrows have them. Each time you observe another sparrow, you test that hypothesis.

Suppose a biology student makes these observations:

1. White blood cells have nuclei, although they may vary in size and shape.
2. Paramecia contain contractile vacuoles, nuclei, and food vacuoles.
3. A textbook illustration of a typical cell shows mitochondria, a nucleus, an endoplasmic reticulum, and other structures.
4. In plant leaf cells, chloroplasts nearly fill the cells, almost hiding the nucleus from view.
5. A description of viruses states that viruses invade their hosts' nuclei.

What hypothesis is the student likely to make concerning the parts of a cell?

Suppose the student continued to make observations. For each of the following observations, state whether it supports or disproves the hypothesis.

human kidney cells: _____

slide of cat brain cells: _____

human red blood cells: _____

organisms in pond water: _____

What might the student do as a result of these hypothesis tests?

Activity 5.3

Making Analogies

An analogy is a way of explaining something by comparing it with something else. In biology, *analogous* means a similarity in function between parts that are dissimilar in structure and origin.

The section on cell structure in your textbook discusses the structure and function of the various organelles found in plant and animal cells. Figure 5.6 in your text illustrates these organelles and gives a brief description of the function of each organelle. The figure on page 45 is an analogy of the inner workings of a cell, in which the artist compares the cell to a factory. Examine these two figures and develop your own analogies for the functions of the organelles of the cell. Include the type of cell (animal, plant, or both) in which the organelle is found, the function of the organelles in the cell, and an analogy that describes its function. The nucleus, for example, is found in both animals and plant cell; it contains the genetic information that directs all cell activities; the nucleus often is referred to as the brain or control center of the cell.

plasma membrane:

nucleus:

cell wall:

mitochondria:

endoplasmic reticulum:

microtubules:

lysosome:

cytosol:

vacuole:

chloroplast:

centrioles:

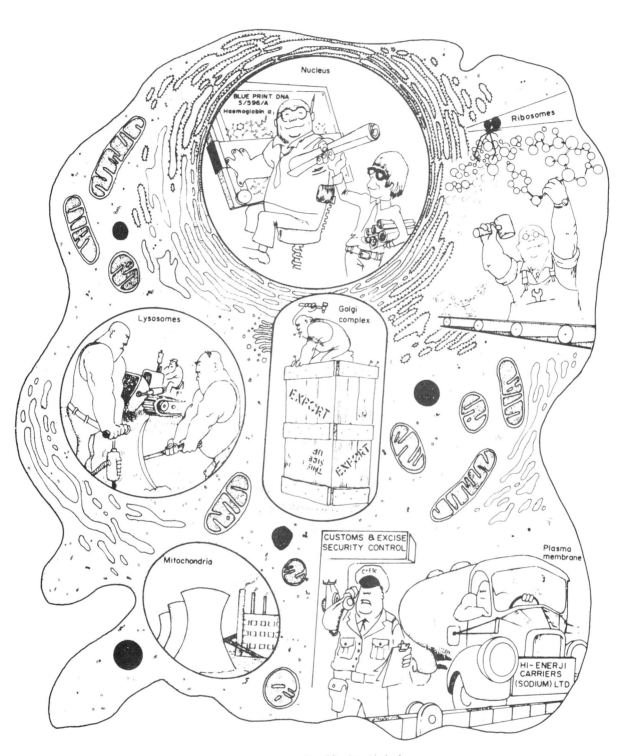

Blackwell Scientific Publications Limited

Activity 5.4

Learning by Questioning

One of the fastest ways to discover whether you understand written material is to try to write a quiz covering it. Review Section 5.10 in your textbook and write a quiz covering the section. Write at least 10 questions. You may wish to provide some sketches to be labeled, or write questions that ask students to draw something. Emphasize questions on topics you think are important and write an answer key.

Close your textbook and try taking the test yourself. When you can answer the questions, exchange your test with another student. Take each other's tests and mark each other's answers. If you disagree on some test items, look them up in your textbook.

Activity 5.5

Diffusion

A typical human body cell contains these proportions of substances:

70% water
1% inorganic ions (such as sodium, potassium, calcium, and magnesium)
15% proteins
7% nucleic acids
3% carbohydrates (2% sugar and 1% starch)
2% lipids
2% other substances

Assume for this activity that water, inorganic ions, sugar, and "other substances" can pass through cell membranes; and that proteins, nucleic acids, starch, and lipids cannot. Assume also that diffusion is based on the percentage composition of substances. For example, an increase in the percentage of CO_2 inside a cell will have the same effect as an increase in the concentration of CO_2, regardless of what happens to other substances in the cell.

Suppose that each column in the table below represents a test tube containing the percentages of substances listed, and the cell at the top is a body cell placed in the test tube. Using colored pencils, draw arrows to show the movements of substances across the cell membrane. Key the pencils to colors referred to in the table.

Percentage Composition of Solution in Test Tube

Substance	Test-tube Number							
	1	2	3	4	5	6	7	8
water (blue arrows)	65	75	68	68	70	73	70	65
inorganic ions (red arrows)	5	1	2	1	1	0.5	1	1
proteins (yellow arrows)	18	10	10	10	13	17	18	15
nucleic acids (green arrows)	7	2	10	7	5	5	7	5
sugar (purple arrows)	3	3	3	3	5	0	1	0
starch (orange arrows)	0	1	2	1	1	2	1	4
lipids (pink arrows)	1	1	3	1	2	2	1	1
other substances (black arrows)	1	7	2	9	3	0.5	1	9

Chapter 6

Activity 6.1

Library Research

An essential part of any scientific investigation, and one of the first steps in the process, is a search of the literature. Once a problem has been defined clearly, the investigator spends many hours in the library searching out as much information about the problem as he or she can find.

This information is valuable in many ways: questions about the problem are answered, research already performed is not duplicated, and the data that is gathered is useful in developing hypotheses. Knowing how to do a search of the literature is a valuable tool, for scientific investigations, but also for any academic endeavor, and even recreational reading.

The information search begins with the card catalog, an index of the library's holdings. It most often is filed in drawers in alphabetical order. In larger libraries, the card catalog may be on microfilm or in computer files.

Each item usually has more than one entry in the card catalog. The title of the source may be on one card, the author on a second card, and the subject on a third card.

 Listed below are three major topics discussed in Chapter 6. Select one of these topics and use the subject headings in a library's card catalog to determine how many books the library has on that particular subject. List the books by title, author, date of publication, city of publication, and publisher.

Topic	Subject headings
Reproduction	Science
	Zoology
	Biology
	Human
Meiosis	Life science
	Fetus
	Sex cells
	Embryology
Fertilization	Pregnancy
	Reproductive system
	Reproduction
	Fertilization
	Cell division
	Genetics
	Fertility
	Birth control

In looking through the card catalog, you may see directions to "see also" other subject headings. If you find other useful subject headings in this way, list them below.

Activity 6.2

Picturing Meiosis

Sometimes in science writing pictures accompany the text, making long explanations unnecessary. At other times, however, pictures are not available or cannot be used. It is important to be able to describe events, objects, and changes accurately and thoroughly with words.

Examine the four drawings of meiotic stages shown below. Describe the cell and its contents at each stage as thoroughly as possible. Try to make the descriptions so complete that someone reading it could picture everything that is shown in the drawings.

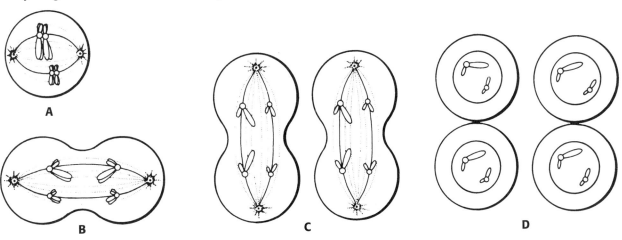

Activity 6.3

The *Drosophila* Life Cycle

The pictures below show the diploid chromosomes of *Drosophila melanogaster*.

In the circles below, draw the chromosomes as they would appear at each stage.

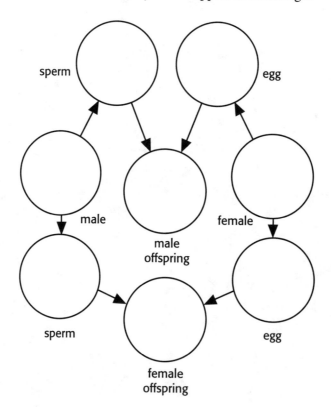

Activity 6.4

Hormonal Activity in the Menstrual Cycle

Hormones are chemical messengers of the body that allow communication between organs and organ systems. The human female reproductive system is controlled by cycles of hormone levels that direct each organ in its role of the menstrual cycle. Review Section 6.7 in your textbook and construct a concept map using the menstrual cycle as the central concept. Include in the map information from the Biology Today on birth control. Some additional terms that should be included are hypothalamus, pituitary, ovary, uterus, estrogen, progesterone, birth control pill, and menopause. Use other concepts as needed to diagram the relationship between hormones and the organ functions they control.

Activity 6.5

Pangenesis

The English biologist Charles Darwin knew nothing of chromosomes or meiosis. These were discovered after his death, when microscopes and staining techniques led to new knowledge of cellular events.

Darwin used the idea of pangenesis, which originated with the Greeks about 2000 years ago, to explain how inheritance takes place. Darwin thought that gemmules (representative particles) came from all parts of the body and entered the gametes. In this way the parents' characteristics were transmitted to the offspring.

The test of any hypothesis is whether the predictions that stem from it are supported by evidence. Make some predictions about the pangenesis hypothesis. For example, what could you predict about the offspring of two brown-eyed parents? What would you predict about the chromosomes in body cells and sex cells? If a man and a woman had lost limbs in accidents before having children, would their children have normal limbs? Compare the predictions with evidence from everyday observations or from studies of cells. If the predictions do not agree with the evidence, the hypothesis is disproved.

Chapter 7

Activity 7.1

Embryonic Development

The early stages of embryonic development are similar for different types of organisms. Look at the three embryos shown below.

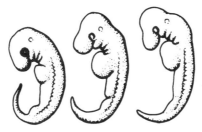

In the space provided, describe what you see in common for all three embryos.

After a short time has passed, the three embryos have changed considerably. Observe the three embryos, and in the space below, describe how each has changed from your first observations.

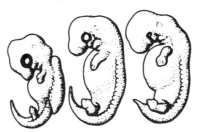

Describe any differences you see between the three embryos. Write your observations in the space below.

Observe below the same embryos after more time has passed. They still are not ready to be born, but they are increasingly mature.

In the space below, describe each of the embryos. If you think you have enough information, identify the type of animal each embryo will become.

Activity 7.2

Understanding Embryonic Development

The SQ3R method is a tool that can help you understand what you read. These are the steps to follow:

1. S: *Skim* the passage quickly.
2. Q: Note the major *questions* you think are answered by the passage. Watch for clues in the printing, such as headings or boldface type. The authors use these to emphasize major ideas or vocabulary.
3. R: *Read* the passage carefully.
4. R: *Recite* the answers to the questions.
5. R: *Review* the passage. Locate any answers you missed earlier.

Try this method on Sections 7.1 through 7.4. These sections deal with one of the most fascinating processes in nature, the development of a multicellular living organism from a single cell. Use the space provided to formulate your questions and to write your responses to the questions you raised.

Questions	Answers

Activity 7.3

Flow Chart of Differentiation

A flow chart helps identify individual steps in a sequence of events. Differentiation begins at fertilization and the formation of the zygote. The process continues as the cell mass grows and tissues begin to form. Review Sections 7.2 and 7.3 in your textbook. Draw a flow chart that illustrates the step-by-step development of an eye lens from a zygote. Briefly describe each step and draw an arrow connecting it to the next step.

Activity 7.4

Development of a Hypothesis

In Activity 7.1 you observed and described the development of three different embryos. At the conclusion of the activity, you guessed what types of animals were represented in the illustrations. You probably guessed correctly for the tortoise. (If you guessed turtle you were very close, but not completely correct.) You also probably guessed correctly for the embryo on the right, a human. What did you guess for the embryo in the center of the illustration? The three stages of embryonic development for the organism are repeated below. Examine the sequence carefully.

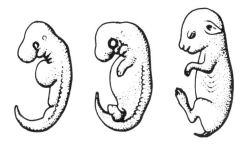

In the space below, write the characteristics that might help you determine what type of organism the embryo is.

A hypothesis is a statement that explains an observation. The observations you listed above are your data. A hypothesis usually is written in an "if . . ., then . . ." format. For example, the following could be a hypothesis about the first organism: "If the first organism in Activity 7.1 has a shell on its back, a shell on its stomach, and a tail, then it probably is a tortoise." In the space below, write a hypothesis about what you think organism 2 might be. Use the "if . . ., then . . ." format.

Biologists formulate hypotheses during much of the time they are conducting research. Carefully formulated hypotheses can be used easily to set up and design an experiment. Find out if your hypothesis is correct by asking your teacher what the organism is. If you were correct, in the space below write the data that were the most important in developing your hypothesis. If you were not correct, go back and examine the organism again. What structures give you the most important clues about what the organism is?

Activity 7.5

Second-Hand Cigarette Smoke

The Surgeon General of the United States has issued several warnings about the dangers associated with cigarette smoking. Smokers are likely to develop lung and other cancers, as well as cardiovascular disease. In addition, smoking by pregnant women has been associated with possible damage to the embryo—among the effects observed are low birth weight and premature birth.

New findings indicate that people who breathe the smoke from other people's cigarettes also are at risk for lung cancer and other respiratory disorders. Deeply imbedded in this issue are the rights of individual persons. Smokers claim the right to smoke, and non-smokers claim they have the right to breathe without being exposed to the hazards of second-hand cigarette smoke.

Take a stand on this issue and support your stand with data. You may use material from your textbook, the library, or from newspapers and magazines. Keep in mind three important questions as you read. First, what are the data? Ask yourself whether the author is presenting data or opinions. Second, what do the data mean? Is the author interpreting the data correctly or twisting the data to make them fit a particular position? Third, what are the professional credentials of the author? Is the author a recognized professional, or someone who is expressing opinions that are not backed up by professional training or academic degrees? Sort through the sources carefully and submit your written report to your teacher.

Activity 8.1

Textbooks are not Perfect

Most people think science textbooks are filled with facts and free of opinions. However, even the authors of science textbooks sometimes are guilty of presenting opinion as fact.

Early in this century, there was a biological and social movement called eugenics. The goal of eugenics was to improve the genetics of the American population, generally by restricting immigration from particular foreign countries and forbidding the marriage of anyone considered genetically inferior. Some of what the eugenicists taught was factual, but much was opinion. For example, they sometimes labeled as "feeble-minded" people who could not speak English or who were physically ill from hookworms or others causes.

George Hunter wrote biology textbooks that were used throughout the country at that time. The following illustration and quote are from a page in his 1914 textbook.

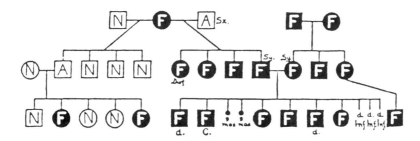

This chart shows that feeble-mindedness is a characteristic sure to be handed down in a family where it exists. The feeble-minded woman at the top left of the chart married twice. The first children from a normal father are all normal, but the other children from an alcoholic father are all feeble-minded. The right-hand side of the chart shows a terrible record of feeble-mindedness. Should feeble-minded people be allowed to marry? (After Davenport.)

The evidence and the moral speak for themselves!

Parasitism and its Cost to Society.   Hundreds of families such as those described above exist today, spreading disease, immorality, and crime to all parts of this country. The cost to society of such families is very severe. Just as certain animals or plants become parasitic on other plants or animals, these families have become parasitic on society. They not only do harm to others by corrupting, stealing, or spreading disease, but they are actually protected and cared for by the state out of public money. Largely for them the poorhouse and the asylum exist. They take from society, but they give nothing in return. They are true parasites.

The Remedy.   If such people were lower animals, we would probably kill them off to prevent them from spreading. Humanity will not allow this, but we do have the remedy of separating the sexes in asylums or other places and in various ways preventing intermarriage and the possibilities of perpetuating such a low and degenerate race. Remedies of this sort have been tried successfully in Europe and are now meeting with success in this country. (Hunter, George, 1914, *A Civic Biology.* New York: American Book Company. Reprinted with permission of D.C. Heath and Company.)

George Hunter. 1914, *A Civic Biology,* New York: American Book Company. Reprinted with permission of D.C. Heath and Company.

What was Hunter advocating? Write a criticism of his argument.

Activity 8.2

Editing a Paragraph

(Adapted from J. D. Watson, 1970, *Molecular Biology of the Gene,* 2nd edition (New York: W. A. Benjamin, Inc.): 187.)

Crossing over occurs at the stage of mitosis where two holologous chromosomes specifically attracteach other to form pairs. The mechanism of this attraction

(*pairing*) remain's a great mystery. It is clearly a very specific process, since it

occurs only among chromosomes containing the same genes. Following the

formation of pairs, both chromosomes occasionally brake at the same point and rejoin

crossways. This allows the formation of recombinant chromosomes

containing some genes derived from the paternal chromosome and some genes derived from

the maternal chromosome. Crossing over greatly increases the

amount of genetic recombination and accept in highly specialized cases, is universally

observed. The frequency of crossing over, however, varies greatly with

the particular specie involved. On the average, one to several crossovers occur every time

chromosomes pair. (Adapted from J. D. Watson, 1970, *Molecular Biology of the Gene,* 2nd

edition (New York: W.A. Benjamin, Inc.): 187).

Activity 8.3

The Lyon Hypothesis

Mutant genes on the X chromosome can cause genetic disorders. For many years, geneticists wrestled with a puzzle involving the X chromosome: why a female who was homozygous for a mutant gene on her two X chromosomes was no more seriously affected by a particular disorder than a male, who had only one such gene on his single X chromosome. In 1961 Mary F. Lyon, a British geneticist, proposed a possible explanation.

In each of the body cells of a normal female, one of the X chromosomes is inactivated. Inactivation occurs in the embryo about 16 days after conception. Either the maternal or the paternal X chromosome is inactivated, apparently by chance, in any given cell. Once inactivation occurs in a cell of a developing embryo, the same X chromosome will be inactivated in all cells that descend from that cell.

Listed below are some data resulting from observations of natural events. Indicate in the space following the data whether it supports or does not support the Lyon Hypothesis.

1. The genes for fur color in mice and cats are carried on the X chromosome.
2. Barr bodies are darkly stained masses that can be seen in the nucleus of certain cells of XX females.
3. Hemophilia is a recessive, X-linked trait that produces severe symptoms in males.

Observations

1. Female mice heterozygous for X-linked genes for coat color had fur made up of patches of the different color. _____

2. Male mice do not exhibit the patchy coat color seen in females. _____

3. Female mice with a single X chromosome (XO) do not show the patches of different color observed in the XX females. _____

4. A Barr body appears to be an X-chromosome that has been wound up on itself. _____

5. An enzyme called G6PD is produced by an X-linked gene in humans. There is no difference between normal females and males in the amount of enzyme produced. _____

6. The amount of G6PD in individuals with 3 or more X chromosomes is the same as that found in XX females. _____

7. The gene responsible for the production of an enzyme called steroid sulfatase is X-linked. Females normally produce twice the amount of this enzyme as males. _____

8. Cells from heterozygous females who have two different alleles for different forms of G6PD on each X chromosome produce only one form of the enzyme. _____

9. Males with Klinefelter syndrome (XXY) have one Barr body in each cell. _____

10. Female carriers of hemophilia may be as severely affected as males, not affected at all, or have symptoms somewhere in between. _____

11. In kangaroos the inactivated X chromosome is always the paternal X. _____

12. Calico cats are heterozygous for a coat color gene (black or yellow). An XXY male cat is calico (black and yellow patches of fur). _____

Activity 8.4

The History of Genetics

As scientists work on their ideas, they are influenced by other scientists who preceded them and those who currently are working. Chapter 8 discusses some of the scientists and the work they contributed to current genetic theory over the past 100 years. As you read Chapter 8, note the scientists mentioned and the important work they did. Construct a time line that illustrates the order of these discoveries and the scientists involved.

Chapter 8 mentions only a few of the events that have led to our current understanding of genetics. Explain how two of these events illustrate how scientists build on previous knowledge.

The changing of ideas in a society takes time. Many scientists are criticized for their new ideas. It can take years before these ideas are accepted by society. List three examples of ideas that may have been rejected when they were first presented to the scientific community.

Think of some of the current research and the hypotheses that you have heard or read about in the media, such as new cancer-causing agents, cancer cures, cholesterol levels, oat bran, and new medical procedures. How might the media play a role in the time it takes for new scientific ideas to be accepted and does society benefit?

Activity 8.5

Considering Ethics

Everyday we make decisions about right and wrong, good and bad. Should a person cheat on a test or make a questionable deduction on his income tax form? Most of these decisions are fairly straight forward. The "rightness" or "wrongness" of the action may be established by society, and it becomes your responsibility to take the consequences of wrong action. If you cheat on a test or an income tax form, you should be prepared to accept the punishment when caught.

Ethics deals with what is good and bad, and with moral duty and obligation. Most of society's ideas about ethics have been established over time, but in the last twenty years, science has progressed so far and so fast, it is difficult for society's ethics to keep up.

 The field of genetic engineering is one arena for debates about the ethics of procedures that now can be performed. Sections 8.13 and 8.14, and Chapter 8's Biology Today in the textbook discuss a few of the developments in genetic engineering and some of the applications that are possible today and in the near future.

Read scientific journals and magazines to find out what new procedures have been developed and the possible applications of these new procedures. Divide into groups and organize arguments for and against the further use or development of these new procedures. Debate the issues in class and develop some ethical guidelines that can be applied to the field of genetic engineering.

Activity 9.1

Evolution Search

Books do not contain all the available information about a topic. The most current information usually is found in journals and periodicals and it is important to know how to locate such information.

Begin a search for current articles about evolution. *The Reader's Guide to Periodical Literature,* found in most libraries, contains references for all articles printed in popular periodicals for each year. This guide usually is found in the reference section of the library, or with the magazines.

Look up the word *evolution,* and locate the listings of articles in periodicals and the list of other related topics. Find and read at least one of the current articles listed. Write down the main ideas and new information presented by the authors.

Activity 9.2

Outlining the Theory of Evolution by Natural Selection

The theory of evolution by natural selection unifies the entire field of biology and it is important to understand the basic premises and ideas that constitute this theory. One way of identifying the important parts of this theory is to outline a section of the text that deals with evolution by natural selection. Section 9.3 provides an overview of how Darwin synthesized the theory of natural selection. Read this section of the text carefully, and then return to the Study Guide. Answer the following questions and use a sheet of notebook paper to write a structured outline of Section 9.3.

What is the title of Section 9.3? Read the first two paragraphs of the section. What is the major, overall idea presented? Write a roman numeral I on the left side of your notebook paper just below the title of Section 9.3. Write the main idea after the roman numeral I. Also, write your answer in the space below.

Read the next three paragraphs. Write a roman numeral II below roman numeral I on your sheet of paper. Next to the roman numeral II, write the main idea covered in the three paragraphs.

Under roman numeral II, add the capital letters A, B, and C, and write the main idea of each of the three paragraphs next to these letters.

Continue the outline through Section 9.4 in the same manner. After you have completed the outline, go back and reread the sections and compare them to your outline. Make any changes that you think are necessary.

Activity 9.3

A Shell Game

The following picture shows the shells of two populations of aquatic mollusks (soft-bodied animals, including snails and slugs.) Each population has a characteristic shell shape. The shells also vary in two other ways. What are they? _____

To keep track of each group of shells, put a number on each one and list the numbers in a column on a piece of paper. Add two more columns to the right of that column and label them A and B. Work with another student. Decide what units to use, and how to measure characteristics A and B.

Fill in the two columns. What was the smallest number in column A? _____

The largest? _____

The smallest number in column B? _____

The largest? _____

Those two pairs of numbers represent the range of each characteristic for the first type of mollusk. Repeat your measurements for the second population. In the space below, describe each population as fully as possible.

Activity 9.4

Grouse: A Species Problem

The greater prairie chicken (*Tympanuchus cupido*) and the sharptail grouse (*Pediocetes phasianellus*) both are members of the order Galliformes. Both birds have interesting reproductive behaviors and have been studied intensively by biologists. The prairie chickens gather on "booming grounds" where the males inflate air sacks in their necks and make the characteristic booming sound. Each male establishes and defends a territory in which he displays independently to females. The sharptail grouse also gather on "booming grounds," but the males display in unison. Both species, however, stomp their feet, lower their heads, and extend their wings in the displays.

The prairie chicken has a fan-shaped tail that is white underneath with a banded pattern on the top. The sharptail has a pointed tail that also is white underneath, but lacks the banded pattern present on the tail of the prairie chicken. The air sac of the prairie chicken is orange. The air sac of the sharptail grouse is purple, and it is smaller. The male prairie chicken erects some feather behind its head during the courtship displays. Whereas the body feathers of the prairie chicken have a distinct barred pattern, the sharptail grouse is rather mottled. Observe the two birds in the following illustration:

Biologists have long hypothesized that these birds represent different genera.

What is the genus of the prairie chicken? _____

What is the genus of the sharptail grouse? _____

The prairie chicken prefers open grassland habitats, and the sharptail grouse prefers open woodlands. Both of these types of habitats occur together in the central part of Wisconsin, and both birds are found in this small area.

What could you do to determine the genetic relationship between these two birds. Write down as many ideas as you can in the space below. State your idea as a hypotheses.

Select one of your hypotheses. In the space below, explain how you would set up an experiment to test your hypothesis. What results would confirm that the birds should be in different genera? What results would indicate the birds are more closely related?

As biologists studied the two birds in Wisconsin, they found a third bird that had a large air sac like the prairie chicken, but it was purple in color like the sharptail grouse. Its tail feathers were intermediate in color and shape between those of the grouse and the prairie chicken, as was the color pattern on its body feathers. Many other characteristics were intermediate between the two birds. How do these data affect the hypothesis that the two birds represent different genera?

These new birds were observed mating among themselves and with both the sharptail grouse and the prairie chicken. Young birds resulted from these matings, so the new birds are not sterile. How do these data affect the hypothesis that the grouse and the prairie chicken represent different genera?

Explain the genetic relationship between the sharptail grouse and the prairie chicken, assuming that both birds have a common ancestor. Explain the processes involved that show how the two birds may have evolved.

In your own words, write a hypothesis that explains the genetic relationship between the sharptail grouse and the prairie chicken, in light of the new evidence. Use the "if . . ., then . . ." format.

Activity 9.5

Finding Fallacies

The paragraph below was taken from a children's science book. Based on your knowledge of evolution, genetics, and scientific methods, write a paragraph with the most vigorous criticism you can of the ideas in the statement. Be alert for fallacies and misconceptions.

> Biologists often try to explain how things work before they have done their experiments. Their explanations are called theories. Theories have to be tested by doing the right experiments. If a scientist proves a theory is correct, the theory becomes a fact. . . . No one has been able to prove that [evolution] is the way life on Earth developed. . . . It will continue to be a theory until someone performs experiments to prove it. (From *Life on Earth: Biology Today,* 1983. New York: Random House.)

Chapter 10

Activity 10.1

How Many Robins?

When studying the evolution of populations, it often is important to calculate the mean (average) value of a characteristic in two or more populations for comparison. Suppose you are studying the brood size (number of young per mating season) of robins. Here is the data you collect:

| Brood | Number of young | | |
	Group A 28° N latitude	Group B 48° N latitude	Group C 62° N latitude
1	4	5	6
2	2	2	6
3	3	6	7
4	4	5	7
5	5	6	6
6	4	6	6
7	4	6	7
8	5	5	6
9	2	4	7
10	3	7	5
11	5	6	7
12	3	6	6
13	3	5	7
14	4	6	5
15	3	5	7
16	4	6	7
17	4	5	5
18	5	6	6
19	3	7	7
20	2	4	6
Mean			
Standard deviation			

Calculate the mean value for each group by adding the brood sizes and dividing by 20. Fill in those spaces in the table.

Notice that the three groups overlap and many values deviate (differ) from the mean for that group. The first brood in Group A consisted of 4 young, whereas the mean was 3.6. The deviation from the mean for the brood was 0.4. You can calculate the deviation from the mean by adding all the separate deviations and averaging them. Calculate the mean deviation for the first three broods for Group A. What is the mean deviation for those 3 broods?

Knowing whether the mean deviation is large or small tells you how the characteristic is distributed. If the mean deviation is large, many values will be far from the mean; if it is small, most values will be very near the mean value.

To calculate the mean deviation for a larger group, use a calculator. Fill in the mean deviation values for the three groups.

Activity 10.2

Taxonomy

Classifying organisms makes communication easier. Scientists developed taxonomy as a tool to identify organisms and given them unique names. Scientists in different countries often had their own names for the organisms they studied and confusion arose when they tried to communicate about their discoveries. A common naming system was needed so scientists could identify a single organism with a fair degree of certainty.

Briefly describe the scientific classification system in terms of Kingdom, Phylum, Class, Order, Family, Genus, and Species. Explain the relationships among the categories.

Similar systems are used in other areas of life to identify specific organisms, people, and objects. Make analogies comparing this man-made scientific classification system to these other man-made classification systems.

1. The classification system used in the library to identify the location of a single book.

2. The classification system in the grocery store that identifies the location of a single food product.

3. The classification system of the postal service that identifies a particular individual to receive a letter.

Scientific classification is not stagnant. It changes to accommodate new discoveries of organisms and newly discovered relationships among organisms. There often is disagreement as to the exact classification of organisms when the relationships are unclear. Explain how each of the analogies you made need to be flexible to accommodate changes.

1. Library—

Grocery store—

Postal service—

Activity 10.3

Why We Have Oxygen

Sections 10.9 and 10.11 of your textbook discuss how changes in organisms have changed the atmosphere. Scientists have studied evidence of oxygen in the rocks and have concluded that the atmosphere's current level of free oxygen was reached about 0.4 billion (400 million) years ago, when large fishes and the first land plants evolved. Oxygen-producing cyanobacteria evolved about 2 billion years ago, and the first living compounds probably were formed about 4 billion years ago, when there was no free oxygen in the atmosphere. Modern photosynthesis may have begun about 1.5 billion years ago.

Draw a rough graph showing the change in free oxygen (from 0 to 100% of the current level) over the last 4 billion years and label important events along the curve.

Activity 10.4

Working as a Team

Work in teams of four and collect a group of at least 20 different pictures of similar animals or plants (such as 20 snakes, 20 conifers, or 20 dogs).

Team member 1 should decide what characteristics to use for classifying the organisms in the pictures and explain the decisions to member 2 only.

Member 2 should make a dichotomous key, based on member 1's instructions, and tell only member 3 how to use the key with the pictures.

Member 3 should follow member 2's instructions to use the key. Based on the results, member 3 should tell member 4 only how the key should be changed.

Member 4 should revise the key and explain the changes, and the reasons for these changes, to member 1.

Activity 11.1

Skimming Paragraphs

One approach to skimming uses paragraphs. In most science writing, the first paragraph states a problem or introduces a topic. The last paragraph sums up the results or reaches a conclusion. Each of the paragraphs in between has a main point, in most cases stated in the first sentence of the paragraph.

Look through Chapter 11 and read the introductory and summary paragraphs thoroughly. Read only the first sentence of each of the remaining paragraphs. Try to understand what the chapter is about from reading only those paragraphs and sentences.

Answer these questions:

What factors may determine the severity of a disease in an individual?

In what ways may a bacterial disease spread through a community?

What beneficial roles do bacteria have in the environment?

Write a few sentences below summarizing what you can remember from skimming the chapter.

Activity 11.2

Archaebacteria

Recently, scientists have discovered some major differences in the ribosomal RNA of a group of bacteria that live in extreme environments. These bacteria also have some characteristics that are similar to prokaryotes,

some that are similar to eukaryotes, and some that are unique. These archaebacteria appear to have changed little from those found in the fossil record.

Review the first three sections of Chapter 11 in your textbook. What characteristics do each of the three groups of archaebacteria have that make them unique?

Thermoacidophiles:

Halophiles:

Methanogens:

Describe the type of environment that would have influenced the development of these organisms.

Where might the environments you described above exist as the earth gradually changed, making possible the development of other organisms? How were archaebacteria able to survive such drastic changes in the atmosphere, soil, and water as plants and animals evolved? Write a paragraph explaining possible answers to these questions.

Activity 11.3

Identifying Eubacteria

In medical laboratories, cultures of bacteria are grown for diagnosis of bacterial diseases. The bacteria are then classified by size, shape, and staining characteristics. The sizes and shapes of some common pathogens are shown below.

Sizes of Common Bacteria

Name of organism	Size of individual cells
Staphylococcus aureus (boils, etc.)	0.8– 1.0 μm in diameter
Streptococcus pyogenes (sore throat, etc.)	0.4– 0.75 μm in diameter
Pneumococcus (pneumonia)	0.8– 1.2 μm in diameter
Gonococcus (gonorrhea)	0.8– 0.6 μm in diameter
Meningococcus (meningitis)	0.8– 0.6 μm in diameter
Corynebacterium diphtheriae (diphtheria)	1.5– 6.5 μm long, 0.3–1.0 μm wide
Mycobacterium tuberculosis (tuberculosis)	2.0– 4.0 μm long, 0.3–0.5 μm wide
Salmonella typhosa (typhoid fever)	1.0– 3.0 μm long, 0.8–1.0 μm wide
Clostridum tetani (tetanus)	2.0– 3.0 μm long, 0.3–0.5 μm wide
Vibrio comma (cholera)	1.0– 2.0 μm long, about 0.4 μm wide
Treponema pallidum (syphilis)	8.0–14.0 μm long, about 0.2 μm wide

Reprinted with permission of MacMillan Publishing Company from *Microbiology* by Kenneth L. Burdon. Copyright ©1968 by MacMillan Publishing Company.

Reprinted with permission of MacMillan Publishing Company from *Microbiology* by Kenneth L. Burdon. Copyright ©1968 by MacMillan Publishing Company.

The Gram stain is a common tool for diagnosis. The stains gentian violet and safranine are used. Some bacteria retain the purple color of gentian violet, others lose the purple color but are stained pink by safranine. The purple bacteria are called Gram-positive, and the pink bacteria are called Gram-negative. This chart shows how some important bacteria react to the Gram stain.

Gram-positive Bacteria		Gram-negative Bacteria	
Organism	**Disease with which associated**	**Organism**	**Disease with which associated**
Staphylococci (all pathogenic species)	— Furunculosis, etc.	Gonococcus	— Gonorrhea
		Meningococcus	— Epidemic meningitis
Streptococci (all important pathogenic species)	— Erysipelas, tonsillitis, scarlet fever, etc.	Salmonella typhosa	— Typhoid fever
		S. paratyphi	— Paratyphoid fever, food poisoning
Pneumococci	— Lobar pneumonia, etc.	Shigella dysenteriae	— Dysentery
		Haemophilus	— Influenza, etc.
Cor. diphtheriae	— Diphtheria	Bordetella pertussis	— Whooping cough
Mycobacterium Tuberculosis	— Tuberculosis	Pseudomonas mallei	— Glanders
		Yersinia pestis	— Plague
Bacillus anthracis	— Anthrax	Fusiform bacilli	— Vincent's angina, etc.
Clostridium tetani	— Tetanus		
C. perfringens	— Wound infection, gas gangrene	Brucella group	— Undulant fever
		Vibrio comma	— Cholera
C. botulinum	— Botulism	Spirochetes	— Syphilis, relapsing fever, etc.
Actinomyces	— Actinomycosis		

Reprinted with permission of MacMillan Publishing Company from *Microbiology* by Kenneth L. Burdon. Copyright ©1968 by MacMillan Publishing Company.

Assume you are a bacteriologist who has just discovered the gonorrhea bacterium in a culture grown from patient's blood. In a short paragraph, describe what the bacteria looked like and how you reached your conclusion.

Activity 11.4

Organizing Information

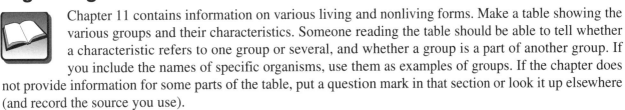

Chapter 11 contains information on various living and nonliving forms. Make a table showing the various groups and their characteristics. Someone reading the table should be able to tell whether a characteristic refers to one group or several, and whether a group is a part of another group. If you include the names of specific organisms, use them as examples of groups. If the chapter does not provide information for some parts of the table, put a question mark in that section or look it up elsewhere (and record the source you use).

Students also might use such characteristics as living or nonliving, role in the nitrogen or oxygen cycle, and so on in their tables.

Activity 11.5

Rice

Rice often is grown in flooded fields, where other plants cannot grow, because of the lack of oxygen around their roots. This is possible because rice plants have an efficient system of air passages that transports air from the leaves to the roots, providing oxygen to the tissues.

Rice plants obtain usable nitrogen from the flooded soil. The nitrogen is fixed by a symbiotic relationship of water fern and cyanobacteria. Bacteria taking part in the nitrogen cycle are found both in the oxidized soil layer and in the reduced (without oxygen) soil layer. They act on the nitrogen compounds that are released as plants above them die and decay.

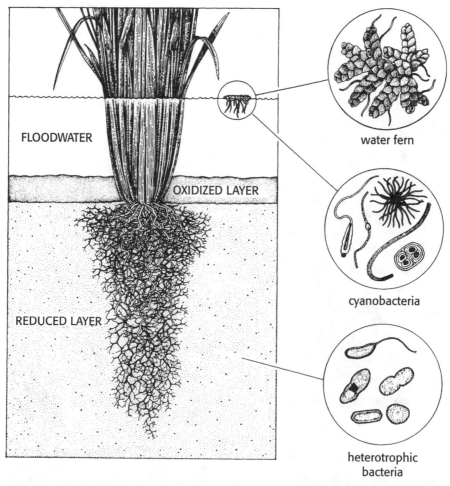

From "Rice" by M. S. Swaminathan, Copyright © January 1984 by *Scientific American Inc.* All rights reserved.

Using the above information, label the diagram to show the cycling of oxygen and nitrogen.

Activity 11.6

Diseases

Bacteria and viruses have many roles in the biosphere, but one role humans focus on is that of pathogens. Which pathogens cause which diseases? How are they transmitted? Who or what do they infect? How can the infection be treated?

Sections 11.6 through 11.11 discuss the relationship of various types of pathogens and the diseases that result from infection. Review these sections and construct a concept map that illustrates these relationships. Consider these subtopics while you organize your ideas.

Bacterial infection

Transmitting agent

Human disease

Toxins

Sexually transmitted diseases

Viral infection

Possible treatment

Plant disease

Retrovirus

AIDS

Chapter 12

Activity 12.1

African Sleeping Sickness

Many protists have parasitic relationships with plants and animals. Most of these relationships involve specific hosts for the parasite. The parasite may have different hosts and free-living stages in its life cycle. Read the following paragraph that describes the life cycle of the flagellated protozoan *Trypanosoma brucei gambienese,* which causes African sleeping sickness in mammals—usually domestic and wild animals such as antelope and buffaloes, and in humans.

The protozoan *Trypanosoma brucei gambienese* has two hosts in its life cycle, an invertebrate and a vertebrate. The invertebrate host, the tsetse fly, is the vector that transmits the infection from one vertebrate to another. The tsetse fly bites an infected vertebrate, such as antelope, and drinks the blood containing the trypanosome. The trypanosome enters the insect's stomach where it adjusts its metabolism to the new energy source—from a vertebrate to an invertebrate. The trypanosome then invades the vector's salivary glands where the protozoan develops into the infectious form. The insect vector then bites a vertebrate, such as a human, and transfers the protozoan into the blood of the new host.

In humans, the trypanosome migrates through the body in the blood, adjusts its metabolism to the new energy source, and reproduces in great numbers. Normally the trypanosome inhabits the blood plasma, cerebrospinal fluid, lymph nodes, and the spleen. If the individual is bitten by another tsetse fly, the trypanosome is transmitted in the blood to the new invertebrate host and the cycle begins again.

The vertebrate host still has the infection and begins to develop symptoms after a one to two week incubation period. Fever, chills, headache, and loss of appetite are the first symptoms to appear. The spleen, liver, and Lymph nodes enlarge and the host becomes weak, anemic, has disturbed vision, and a low pulse rate. As the nervous system is invaded, the person readily falls asleep. Coma and death often result after a few years. During this time, the human host is able to transfer the trypanosome to all the tsetse flies that bite him, therefore, spreading the infection.

Summarize the sequence of events in the life cycle of *Trypanosoma brucei gambiense* in a flowchart. Use arrows to indicate the sequence. Since this is a life cycle, arrows may be used to indicate the return to the initial steps in the sequence.

Activity 12.2

Using Tables to Summarize

The kingdom Protista is a catchall of organisms and it is difficult to characterize its members. The following chart will help summarize the major groups of protists. Review Sections 12.1 through 12.9 in your textbook and complete the table using characteristics that apply to each group.

	algae				protozoa				slime molds
	green	diatoms	brown	red	flagellates	sarcodines	sporozoans	ciliates	plasmodium
unicellular									
colonial									
multicellular									
specialization									
motility									
nonmotile									
flagellated									
ciliated									
pseudopods									
gliding									
autotrophic									
heterotrophic									
predator									
decomposer									
reproduction									
asexual									
sexual									
spore-forming									
lifestyle									
free living									
symbiotic									
parasitic									

Activity 12.3

Finding the Nucleus of an Idea

The word "nucleus" can be used in many connections, but it always refers to something central. Most cells have nuclei, and so do most paragraphs. Read the following paragraphs and try to pick out the central idea in each one.

(1) I was raised in the belief that organelles were obscure little engines inside my cells, owned and operated by me or my cellular delegates, private, submicroscopic bits of my intelligent flesh. Now, it appears, some of them, and the most important ones at that, are total strangers.

(2) The evidence is strong, and direct. The membranes lining the inner compartment of mitochondria are unlike other animal cell membranes, and resemble most closely the membranes of bacteria. The DNA of mitochondria is qualitatively different from the DNA of animal cell nuclei and strikingly similar to bacterial DNA; moreover, like microbial DNA, it is closely associated with membranes. The RNA of mitochondria matches the organelles' DNA, but not that of the nucleus. The ribosomes inside the mitochondria are similar to bacterial ribosomes, and different from animal ribosomes. The mitochondria do not arise *de novo* in cells; they are always there, replicating on their own, independently of the replication of the cell. They travel down from egg to newborn; a few come in with the sperm, but most are maternal passengers.

(3) The chloroplasts in all plants are, similarly, independent and self-replicating lodgers, with their own DNA and RNA and ribosomes. In structure and pigment content they are the images of prokaryotic blue-green algae. It has recently been reported that the nucleic acid of chloroplasts is, in fact, homologous with that of certain photosynthetic microorganisms.

(4) There may be more. It has been suggested that flagella and cilia were once spirochetes that joined up with the other prokaryotes when nucleated cells were being pieced together. The centrioles and basal bodies are believed in some quarters to be semiautonomous organisms with their own separate genomes. Perhaps there are others, still unrecognized.

(5) I only hope I can retain title to my nuclei.

From "Organelles as Organisms" from *The Lives of a Cell* by Lewis Thomas. Copyright © 1972 by the Massachusetts Medical Society. Originally published in *The New England Journal of Medicine.* Reprinted by permission of Viking-Penguin Inc.

What is the main idea in paragraph 1 ?

In paragraph 2?

In paragraph 3?

In paragraph 4?

In paragraph 5?

Activity 12.4

Reading to Find Hypotheses

Read the following paragraphs and identify the hypothesis. Information in Chapter 12 of your textbook may be helpful.

Bordeaux mixture, a chemical mixture used for controlling plant diseases, was developed by a French botany professor, Alexis Millardet. Millardet was working with some grape growers in the 19th century.

The growers had imported grape seedlings from the United States and were growing them with some success. However, they were plagued with two problems: grape thieves and a plant disease. The disease, called downy mildew of grapes, was spreading from the American seedlings throughout the vineyards. In an effort to curb the grape thieves, the growers put a foul-tasting mixture of copper sulfate and lime on the grapes. Millardet noticed that the grapes having that mixture on them did not become diseased.

What was Millardet's hypothesis?

Activity 12.5

Whodunit?

Solve this murder mystery:

In 1932, a healthy person traveling in Wales died in a locked hotel room. Not only was the door locked, but the windows were nailed shut.

For some reason, the hotel maids were superstitious about the room and insisted that the room itself was capable of killing people. While skeptical of that idea, the police were baffled as to the cause of death, and they began an investigation of the room itself.

It looked like a room for murder—dark and damp, it was papered with a dark green velvety paper that added to the gloom. The wallpaper looked normal, but the police found that it contained arsenic. However, the victim scarcely was likely to have eaten the wallpaper. Only if converted to a gas might it have poisoned the victim; and even then, a high concentration of the toxic gas would have been needed.

In reading up on arsenic compounds, one of the police detectives discovered that a fungus, *Scopulariopis brevicaulis,* can convert arsenic to a gas, trimethylarsine.

Can you explain the murder?

Activity 13.1

Finding Information on Plants

Before reading Chapter 13, look through the chapter, noting the pictures especially. Find a plant or group of plants that interests you—perhaps one that grows in your area, or that you have seen when traveling or living elsewhere.

Go to the library and gather enough information on that plant to give a five-minute class report on it. Find out such things as what unusual characteristics it has, where it lives, what animals eat it, how humans use it, and so on. Write a bibliography of the references you use.

Activity 13.2

Reporting on Plants

Use your notes from the last activity to prepare a five-minute class report on your plant. Use only a few notes for your talk to help you talk naturally rather than reading every word. If possible, bring the plant or a picture or slide to illustrate the talk. Give your teacher a copy of the bibliography you prepared.

Activity 13.3

Tentative Explanations

Chlorophyll may be found in many forms that have different characteristics. Two of the most important forms are chlorophyll *a* and chlorophyll *b*. Both types may be involved in photosynthesis, but chlorophyll *b* absorbs more light in blue wavelengths than does chlorophyll *a*. Thus, chlorophyll *b* is more efficient in allowing photosynthesis in shady places.

The following diagram shows how the plants changed on an abandoned mid-latitude farm during 150 years.

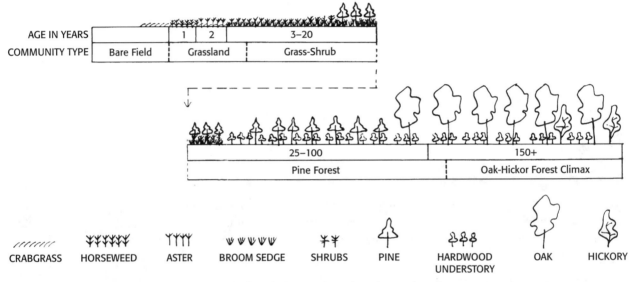

Copyright © 1967 by Rand McNally and Company from *Plant Pigments* by Robert W. Stegner.

Note that oak seedlings can grow beneath pine trees, but pine seedlings are not found under oak trees. Eventually, as a result, the area became a forest of oaks and other hardwoods, with few or no pines.

Based on the above information, write a hypothesis about the chlorophyll in oaks and pines.

Activity 13.4

Primitive Plants

The plant shown below is a horsetail, or *Equisetum.* The stems of horsetails are hollow, jointed branches of underground stems. In most areas they are less than 2 m high, although in the tropics one species grows as high as 12 m.

X1

Equisetum

In the Carboniferous Era, giant plants related to modern horsetails grew in many places and they thrived in the warm, swampy environment. One example is *Calamites,* shown below.

X1/151

Calamites

As the earth became drier, the swamps receded, and the habitat of the giant plants disappeared. The plants themselves also disappeared.

Modern horsetails have the same requirements as their ancient relatives, yet they thrive in many areas. In the space below, write your explanations of how you think this is possible.

Activity 13.5

Coevolution

Part I

Review Section 13.9 in your textbook. In this section, the coevolution of the hummingbird and the red flower it pollinates are discussed. How are the characteristics of the red flower and large amounts of nectar beneficial both to the plant and to the hummingbird?

How is the shape of the flower, with the nectar at the bottom of a long tube, beneficial to both species?

What selective pressures may have helped shape the characteristics of the nectar tube in the plant and the long tongue of the hummingbird?

Part II

Carefully read the description of the three flowering plants and study the drawings of the flowers.

Wand Lily

◆ This simple dish-shaped flower is easy to land on.
◆ The nectar and pollen are easy to find.
◆ It is visited both by short-tongued and long-tongued insects.
◆ It blooms July through August.

Wavyleaf Thistle

◆ It is a more complex tube-flower.
◆ The nectar is found at the base of the tube, selecting for long-tongued insects.
◆ Insects land on the flower head.
◆ The pollen is easy to find; it is presented to the insects on the developing pistil.
◆ It blooms mid-June through mid-August.

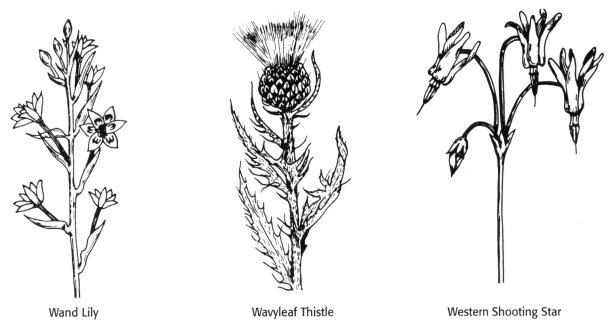

| Wand Lily | Wavyleaf Thistle | Western Shooting Star |

Rocky Mountain Plants, 1969, Ruth Ashton Nelson. Drawings by Dorothy Leake and Ruth Ashton Nelson, reprinted by permission.

Western Shooting Star

◆ It is a most complex flower.
◆ It is an inverted tube flower with no landing platform for insects.
◆ The pollen and nectar are both hidden.
◆ Insects must hover upside-down and separate the stamens with their long tongues.
◆ The nectar is found at the base of the stamen tube.
◆ Pollen falls onto the underside of the insect when it vibrates its wings while hovering upside-down.

Three likely pollinators are a fly, a butterfly, and a bumble bee. You may need to look up the general characteristics of each of these insects in an insect guide.

1. What characteristics make the fly unique?

From which of these flowers would the fly be able to collect the nectar and pollen, resulting in pollination?

Is the fly specially adapted to collect pollen from any of these flowers that the bee or butterfly might not be able to visit?

2. What characteristics does a butterfly have that make it unique?

Which of these flowers would the butterfly be able to visit successfully?

Is the butterfly especially adapted to collect pollen from any of these flowers that the fly or bee might not be able to visit?

3. What characteristics does the bumble bee have that make it unique?

Which of these flowers would the bumble bee be able to visit successfully?

Is the bumble bee especially adapted to collect pollen from any of these flowers that the fly or butterfly could not visit successfully?

4. Which flower is accessible to the most variety of pollinators?

What advantages does this plant and the pollinators have?

What disadvantages does this plant and the pollinators have?

5. Which flower is accessible to the fewest variety of pollinators?

What advantages does this plant and the pollinator have?

What disadvantages does this plant and pollinator have?

6. Discuss what selective pressures may have enhanced the coevolution of the Western Shooting Star and bumble bee so that they became dependent each other.

Activity 14.1

Indoor Birdwatching

Birdwatchers use many characteristics for identifying birds. What might they use for identifying birds at a distance?

What characteristics would be useful for identifying birds at close range?

The following illustrations show a variety of sparrows found in the United States (excluding Hawaii and Alaska) west of about 100° longitude.

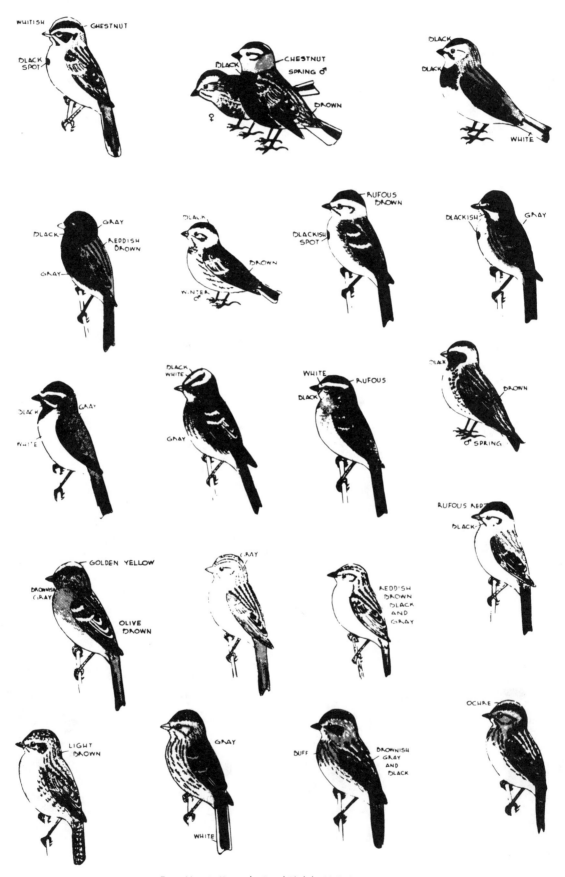

From *How to Know the Land Birds* by H. E. Jaques.

In what states would you be most likely to see these sparrows?

Use the characteristics shown in the pictures to construct a dichotomous key for identifying these species.

Activity 14.2

A Whale of a Tale

In the passage below, look for misused terms, repetitive words or phrases, bad sentence construction, and other mistakes that interfere with clarity. Number each mistake with a colored pencil. Correct the mistake or make whatever changes are needed to increase clarity.

Because of their fishlike shape and aquatic habitat, whales were once thought to be large fish. Actually they are mammals and, as with humans, they have lungs and give birth to live young, and they nurse thier young. The few hairs to be found on a whale are on its muzzle. Whales probably descended from terrestrial animals that adapted to aquatic life, which might explain there resemblance to fish.

Whales are well-adapted to their life in the world's oceans. Beneath their skin they have a layer of blubber, or fat. It provides a source of stored energy and helps conserve heat. Their very, very streamlined bodies enable each and every species of whale to swim and dive in all the oceans of the world. The whale cannot move it's head separately from its body, because the neck vertebrae are compressed into a short area.

A whale's flippers are small in proportion to its body, making the tail the main organ for locomotion.

Their lungs are huge, even in proportion to their body size. Both the lungs and the heart are adapted to those conditions and help the whales take in and conserve large amounts of oxygen. Whales need powerful lungs and circulatory systems, because they dive to great depths and may stay down for as long as an hour.

Activity 14.3

Seeing From All Sides

When studying animals, biologists often make use of prepared slides. Tissues are cut at various angles, and they are studied as cross sections, longitudinal sections, and so on.

The picture below shows the human kidney from the side.

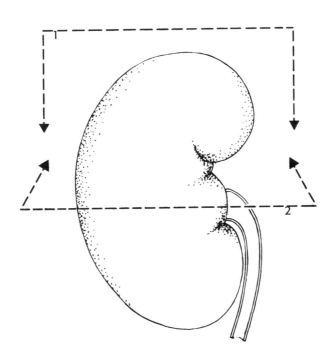

This picture shows what a section of the kidney would look like if cut in the plane indicated by the dotted lines (1) at the top.

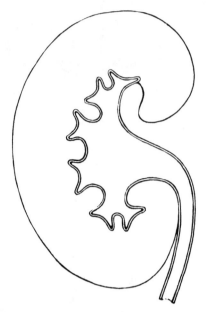

In the space below, draw what the section indicated by the dotted lines (2) would look like.

Concept Mapping Diversity and Adaptation in Animals

Much of evolution is a result of random mutations that allow members of a species to adapt to changes in their environment better than other members. How might this affect the population as a whole?

What would happen to the frequency of this mutant gene if the associated trait was beneficial to the population?

Recall what you learned in Chapter 3 about populations and ecosystems. What are some of the necessary elements in an ecosystem for which members compete?

Mutations and competition for scarce resources are the sources of the selective pressure that have influenced the evolution of animals.

 Review Sections 14.3 through 14.7 and list the traits or characteristics for each animal group that makes it uniquely adapted to survive in a changing environment.

Invertebrates

Sponges

Worms

Insects

Vertebrates

Fish

Amphibians

Reptiles

Birds

Mammals

Using this information, construct a concept map that illustrates the evolution of the animal kingdom. The characteristics should be part of the idea and the selective pressure that enhanced its usefulness should be the connecting term. An example might be amphibians developed the characteristic of breathing air. The selective pressure was the need to move to the land where there was less competition.

Activity 14.5

On Being Thick-Skinned

An insect or other arthropod is covered with a cuticle, which lines many body cavities, forms the exoskeleton, and may form the wings. In addition to supporting the animal, the cuticle provides an anchoring place for muscles and provides a barrier between the animal and its environment. One of its roles as a barrier is to protect the animal from water loss. The chitin, of which the cuticle is made, is a polysaccharide similar to cellulose.

Write a paragraph describing insect cuticles by comparing them to the cell walls of plant cells. Begin the paragraph with a topic sentence that states the main idea.

Chapter 15

Activity 15.1

Some Medical Detective Work

The liver secretes bile, a mixture of substances that includes pigments and salts. The salts are responsible for breaking down fat droplets in the small intestine. Many blood vessels pass through the liver, and a large vein from the liver travels in the direction of the heart.

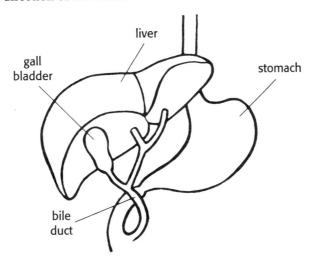

A patient had a severe case of jaundice (yellowing of the skin and, especially, in the whites of the eyes). The physician ruled out several causes of this symptom, and eventually discovered that the patient had a large tumor in the upper portion of the small intestine. Explain how the tumor could cause jaundice.

Activity 15.2

Cellular Respiration

Review Sections 15.5 through 15.9 and outline the processes of cellular respiration. Include the steps of glycolysis, Krebs cycle, and electron transport system. Use the section titles as your roman numeral headings.

Activity 15.3

Using Tables of Data

Tables are a way of presenting information in a form that takes up little space. Tables also present the information in a form that is easier to use for comparison and to discover trends.

One place tables are found is on the labels of food packages. One table on the nutrition information label lists the amount of each type of biological molecule found in the food. There also is a table that lists the percentage of the daily recommended allowance of vitamins and minerals found in the food.

Why would manufacturers use a table to list this information instead of writing it in paragraph form as they do the ingredients?

The following table gives the nutritional information found on the labels of two types of cereal and a can of soup. As you study the table, notice that not all labels contain the same amount of information.

Why do you think this information is not consistent on all food packaging?

Nutrition Information from Food Labels

Component	Cereal A	w/milk	Cereal B	w/milk	Soup
Serving size	1oz, 1c	1/2c	1oz, 1/2c	1/2c	4oz, 1/2c
Protein, g	2	6	3	7	6
Carbohydrates, g	25	31	20	26	21
Fats, g	0	0	4	4	3
Saturated, g	0	0	3	3	-
Unsaturated, g	0	0	1	1	-
Cholesterol, mg	0	0	0	0	-
Calories	110	150	110	150	130
Sodium, mg	290	350	150	210	-
Potassium, mg	35	235	160	360	-

Percentage of US Recommended Daily Allowance					
Protein	2	10	4	15	8
Vitamin A	15	20	15	20	6
Vitamin C	25	25	25	25	6
Thiamin	25	30	25	30	*
Riboflavin	25	35	25	35	2
Niacin	25	25	25	25	2
Calcium	*	15	*	15	4
Iron	10	10	10	10	8
Vitamin B6	25	25	25	25	-
Vitamin D	10	25	10	25	-
Folic Acid	25	25	25	25	-
Phosphorus	4	15	10	20	-
Magnesium	2	6	10	15	-
Zinc	2	6	10	15	-
Copper	2	4	10	10	-

- information not given on label
* percentage less than 2%

Look at the information in the upper portion of the table under the two cereals, A and B. Compare serving size, calories, protein, carbohydrate, fat, cholesterol, sodium, and potassium content.

1. Which nutrients are healthier in higher quantities?

Which nutrients are healthier in lower quantities?

Milk has which components and in what amounts found in the one half cup served with the cereals?

Does one cereal appear to be more nutritious?

Which components does it appear to contain healthier amounts?

2. A calorie is a unit of measure for energy. The cereals have how many calories available as energy for the body to use?

What type of food molecule would you expect to contain the majority of these calories?

3. Why might the sodium and potassium content be important information to some consumers?

4. Look at the bottom half of the table under the two cereals. These figures are percentages of the daily requirements of nutrients present in the food that are necessary for a healthy diet. How do the two cereals compare?

What are some of the differences?

5. How does the soup compare to the cereals in nutrient content?

What are some of the differences?

6. Which of these foods would be the most nutritional snack? Why?

Look at the labels found on some other types of food. What nutrients might be found in high amounts in typical American snack food?

Activity 15.4

Animal Protein: How Essential?

For each kilogram of your ideal body weight, you need about 0.5–0.8 of protein per day. For example, a person of your age weighing 54 kg would need an average of 35.1 g of protein per day. You can obtain this protein from a variety of foods. For example, 85 g of cooked beef or chicken yields 24 g of protein, 50 g of broccoli yields 3 g of protein, 6 g of peanut butter yields 4 g of protein, and 50 g of cooked beans yields 8 g of protein.

For many years, athletes were fed huge amounts of steak, eggs, and other high-protein foods from animal sources because people thought the animal protein would help the athlete build muscle tissue and have stamina. Recently, however, studies indicate that most proteins from animal sources also contain large amounts of cholesterol, which may contribute to cardiovascular diseases. Athletes today eat less protein from animal sources; some are vegetarians. They appear healthy, are good athletes, and have low levels of cholesterol in their blood.

Protein from plants often is incomplete—some of the essential amino acids are missing. Corn, for example, has little of the essential amino acid lysine. Beans have lysine, but little methionine. If you tried to live on just corn or beans, eventually you would suffer from a protein deficiency.

State a hypothesis about the need for animal protein. List any assumptions on which your hypothesis is based. For example, you might assume that only muscle tissue can be used to make muscle tissue. From the information above and that presented in Chapter 15, refute the hypothesis by making at least one prediction from your hypothesis and show that the prediction does not agree with the evidence.

Hypothesis:

Assumptions:

Prediction(s):

Activity 15.5

Fad Diets

Americans have an attraction to fad diets. The possibility of a quick fix to a health problem or the possibility of improved health draws thousands of people to products advertised on television, radio, or in the mail. The following paragraphs are from *The Surgeon General's Report on Nutrition and Health.*

Read the following paragraphs closely. There are ten mistakes in spelling, grammar, or usage. Circle the mistake and suggest a way of improving the sentence.

Contemporary food fads often make one or more of the following claims, none of which are substantiated by available scientific evidence:

- ◆ Some foods have magical, life-promoting propertys.
- ◆ Modern foods are grown on depleted soil, are overprocessed, and therefore, cannot provide good nutrition.
- ◆ Food supplements are always necessary to ensure well nutrition, and megadoses of nutrients provide "supernutrition."

Food faddism is a dietary practice based upon an exaggerated belief in the effects of food or nutrition on health and disease. Food fads derive from three beliefs: (1) that special attribute's of a particular food may cure disease, (2) that certain foods should be eliminated from the diet because they are harmful, and (3) that certain food convey special health benefits. Unlike more transitory fads, many key concepts associated with food faddism persist or reapear periodically. Food faddists are those who follow a particular nutritional practice with excessive zeal and whose claims for it's benefits are substantially more than science has substantiated. Food praised as beneficial, in most instances, such as special products or vitamin supplements, are not as good as faddists claim, and those foods condemned as harmful, such as white flower or sugar, are not as bad.

Nutrient supplements are usually safe in amounts corresponding to the RDA, but the RDA's are already set to provide maximum benefit consistent with safety. Thus, there is no reason to think that larger doses will improve health in already healthy people, and excess intake can be harmful. Megadose intakes (often defined as 10 times or more the recommended levels) can have seriously harmful affects. Excessively restrictive dietary practices can also induce serious medical problems or even death. popular weight reduction products, for example provide very low daily calorie intakes. Because such products have been associated with the deaths of some young women, the FDA now requires warnings on all labels to alert consumers of the potential of such products. Commerical interests have capitolized on a heightened public awareness of nutrition and health issues, but much of the public cannot evaluate the validity of available weight reduction schemes, supplements, and services. (Adapted from U.S. Department of Health and Human Services, 1988, *The Surgeon General's Report on Nutrition and Health.* Public Health Services, pp. 695–705.)

Activity 16.1

Approaching Chapter 16

You used the SQ3R technique in Activity 7.2. Use it again now to prepare for studying Chapter 16. Write your questions and answers in the space below.

Questions	Answers

Activity 16.2

Writing a Short Bibliography

At the end of nearly every science paper is a bibliography. (It may be called by other names—References, Sources, Literature Cited, and so on.) A bibliography is a list of books and papers used as references by the author. It is the last thing written, but the first thing used by the writer.

Go to the library and find five books or articles on AIDS, autoimmune diseases, or other topics dealing with immunity. Write a bibliography of those five references using the format shown—the one recommended by the *American National Standard for Bibliographic References*. Follow the format below for articles:

Author's last name, author's first name or initials. Year of publication. Title of article. Name of journal or magazine, volume, pages.

For Example:

Mills, J. and Masur, H. 1990. AIDS-Related Infections. *Scientific American* 263: 50–59.

Follow this format for books:

Author's last name, author's first name or initials. Year of publication. Title of book. Place of publication: publisher.

For Example:

Hood, L., Weissman, I., and Wood, W. 1978. *Immunology.* Menlo Park, CA: The Benjamin/Cummings Publishing Co., Inc.

Usually a bibliography is arranged alphabetically, according to the authors' last names.

Although you can find references anywhere, be careful to verify them by checking the original article or book. Bibliographies often contain mistakes, and sometimes a library card catalog can be wrong.

Activity 16.3

Vitamin C

Kidney stones are painful accumulations of mineral compounds in the collecting ducts of the kidney. The cause of kidney stones is not understood completely, but they often are associated with a high intake of sulfa drugs, vitamin C, or other compounds.

Vitamin C (ascorbic acid) is important for maintaining the strength of the skin and mucous membranes. The vitamin C molecule itself is composed of atoms of oxygen, hydrogen, and carbon. It contributes to the formation of compounds that also may contain nitrogen and minerals.

Persons who are deficient in vitamin C may have such symptoms as bleeding gums and bruised skin. The vitamin is available as a water-soluble crystalline substance, usually taken in tablet form, and often is considered harmless because it is filtered out of the blood and excreted. Thus the vitamin cannot build up to toxic levels in the blood or body.

From what you know of several human body systems, explain how vitamin C enters the body, is distributed, and is excreted.

Activity 16.4

Temperature Regulation Concept Map

Humans and other animals use several internal mechanisms for regulating body temperature. Maintaining a constant body temperature has several advantages. Review Sections 16.12 through 16.14 in your textbook and construct a concept map that illustrates the advantages of constant body temperature and mechanisms used to maintain it. Include ideas regarding body activities that affect body temperature.

Activity 16.5

Reviewing

Now that you have read Chapter 16, you probably will find your ideas about what is in the chapter are somewhat different from when you used the SQ3R technique. To prepare for your next test, write a series of questions covering the chapter in the space below. For each heading listed, write at least three questions that you think are important.

Circulation

Immunity

Blood

Respiration

Excretion

Homeostasis

Reproduction

Activity 17.1

Skimming for Meaning

Chapter 17 discusses the effects of drugs on the human nervous system, which bring about changes in surface behavior. Read the following passage quickly, then close the book and write down as much as you can remember of the passage on a piece of paper.

Brain Effects of Marijuana

Immediate effects on the brain are the least controversial and best defined of marijuana's hazards. Like alcohol, marijuana is intoxicating. A marijuana high interferes with memory, learning, speech, reading comprehension, arithmetic problem-solving, and the ability to think. Driving skills are impaired, as is general intellectual performance. Long-term intellectual effects are not known.

Some researchers have described what they called motivational syndrome among young marijuana smokers, who, with frequent use of the drug, tend to lose interest in school, friends, and sexual intercourse. However, it is not known whether marijuana use is a direct cause or merely one symptom of a general underlying problem. Persistent brain abnormalities and changes in emotion and behavior have been demonstrated in monkeys given large doses of marijuana.

Like alcohol, marijuana interferes with psychomotor functions such as reaction time, coordination, visual perception, and other skills important for driving and operating machinery safely. Actual tests of marijuana-intoxicated drivers have clearly shown that their driving is impaired, yet they tend to think they are driving better than usual. In several surveys 60 to 80 percent of marijuana users said they sometimes drive while high.

Marijuana is not physically addicting, but people can become psychologically hooked on the drug. Marijuana may aggravate existing emotional problems. The most common adverse emotional effect is an acute panic reaction, in which the user may become terrified and paranoid and require hospital treatment. In 1978 some 10,000 people were treated in hospital emergency rooms for adverse marijuana reactions. (From Jane Brody, 1982, *The New York Times Guide to Personal Health* (New York: New York Times Books)).

The passage above provides information on marijuana's effects on the brain. What other organs are affected by the drug?

Activity 17.2

Making a Flow Chart

The transmission of a nerve impulse may appear instantaneous, but it involves a complex sequence of events. Review Section 17.4 on the transmission on nerve impulses. Construct a flow chart that depicts, step by step, the events from a receptor neuron in the finger receiving a stimulus (a pin prick), the impulse traveling to the central nervous system and being interpreted, and the response being sent to the motor neuron and effector muscle (causing the finger to move away).

Activity 17.3

A Pattern of Behavior

The marine snail lays its eggs according to a rigid pattern of behavior. The actions are determined by a set of genes.

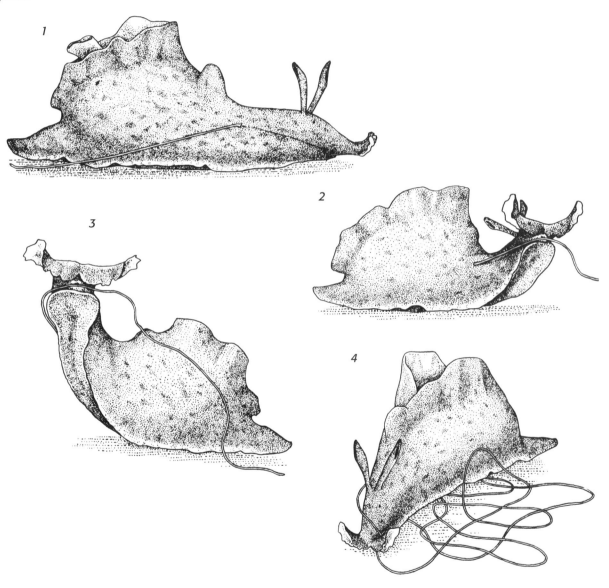

1. The muscles of the reproductive tract contract, expelling a string of egg cases.

2. The animal takes the string in its mouth.

3. At the same time, it waves its head, which pulls the rest of the string out of the reproductive duct.

4. When an entire string has emerged, the snail attaches it to a solid surface.

These behaviors can be induced in the laboratory by injecting the snail with peptides. The peptides ordinarily are produced by bag cells in the snail's brain—abdominal ganglion. As the peptides are produced, they act as neurotransmitters and affect neurons within the brain.

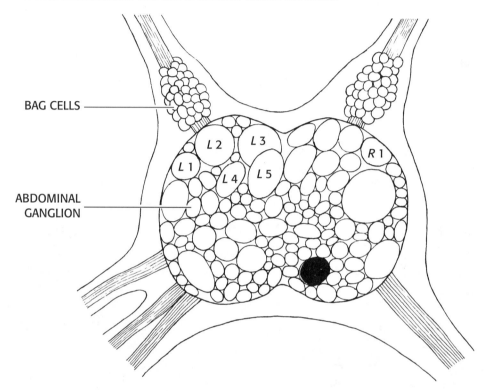

Gene A provides the information that causes bag cells to produce peptide A. Peptide A excites the neurons labeled L1 and R1 in the drawing. Similarly, gene B brings about the production of peptide B, which inhibits neurons L2, L3, L4, and L6. Gene C controls peptide C, exciting neuron R15; peptide C also acts as a hormone that causes contraction of the reproductive duct. Exactly how the peptides control the rest of the behavior pattern is not known.

Make some assumptions about the control of the behavior pattern and assume a mutation in the snail's DNA adds a new step to the behavior pattern, or omits a step. In two or three paragraphs, list your assumptions and explain what must happen between the mutation and any change in the snail's behavior.

Activity 17.4

The Control of Blood Sugar Level

Use your textbook as a source of information to model how the endocrine system controls the level of glucose in the blood. Draw your finished model in the space below.

A tumor in the hypothalamus causes a decrease in secretion of its hormones. Use your model to predict the effect of that decrease on blood glucose level.

What would be the secondary effects of that change on the body?

Activity 17.5

Cocaine

Cocaine affects the user through its chemical action on sensory neurons. Cocaine molecules block the channels through which sodium ions move in and out of axons.

Explain why this has an anesthetic effect on the cocaine user.

Cocaine also affects surface behavior for chemical reasons. The cocaine molecules move into the synapses of neurons in the sympathetic nervous system. The axons of these neurons, like those of other neurons, release neurotransmitters such as norepinephrine into the synapses. Ordinarily, the neurotransmitters are reabsorbed into the axon. Cocaine interferes with that reabsorption.

Explain what that interference would do to transmission of an impulse across the synapse, and what the effects on the person's muscles and glands would be. (You can refer to your textbook as well as to the art below.)

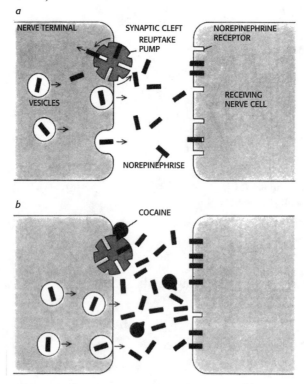

From "Cocaine" by Craig Van Dyke and Robert Byck. Copyright © 1982 by *Scientific American, Inc.* All rights reserved.

Chapter 18

Activity 18.1

Yarrow Growth

Examine the pictures and graph below:

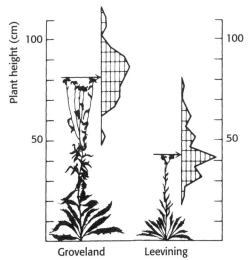

Groveland   Leevining

J. Clausen, D. D. Keck, and W. M. Hiesey, *Experimental Studies on the Nature of Species. 111. Environmental Responses of Climatic Races of Achillea,* Carnegie Institute of Washington publication 581(1948).

Each plant in the picture represents 60 specimens of yarrow plants, Achillea lanulosa. The plants all were grown under controlled conditions at Stanford University, but differences between the Groveland and Leevining populations resulted in different heights. The arrows across the tops of the plants represent the mean height of each population, as shown on the Y axis.

What was the mean height of plants in the Groveland population? _____

What was the mean height of plants in the Leevining population? _____

The graph at the right of each plant picture shows the range of heights in each population. (These graphs may be easier to understand if you turn the page 90° counterclockwise.) Each square represents 2 individuals of a given height. For example, for the Groveland population there were about 8 individuals of 70 cm height. What height did the greatest number of individuals in the Leevining population have?

_____

The range of height for the Leevining population was about 19 cm to about 82 cm. What was the total range of height for the Groveland population?

___

If the two populations had been plotted on a single graph, what would the curve have looked like?

Activity 18.2

Transport Analogies

Stems have two major functions in the plant: support and conduction. Review the section on stems and conduction in your textbook. The inside of the stem contains vascular tissue that transports nutrients and water to all parts of the plant. Think of how it might do this. Do nutrients from the leaves travel in the same vascular tissue as water from the roots?

One way to understand this conduction system is to compare it to similar systems more familiar to you. Write a paragraph using an analogy to compare these similarities to each of the following transport systems.

The Freeway System

The Human Circulatory System

Activity 18.3

Sales Resistance

A farmer's apple orchard was not doing well; many of the leaves were small. In hopes of improving the orchard, the farmer asked a traveling salesman to do some soil tests and recommend a product that might help the trees. The salesman reported that the soil was deficient in zinc, and that the farmer should purchase an additive containing zinc. The additive, incidentally, was quite expensive.

    Based on Table 18.2 in your textbook, what advice would you give the farmer?

Activity 18.4

Plant Growth

The following paragraph is taken from an article about the development of plant leaves.

Plant development is quite unlike animal development. With most animals the formation of new organs is confined to the earliest phases of embryonic growth. With plants the process is a continuous one. New organs arise from perpetually embryonic growth centers: undifferentiated tissues consisting of cells capable of transformation into a variety of plant organs. The growth centers, at the extremities of the plant, are embryonic tissue at the apex of the root and shoot. Such tissue is termed the apical meristem. (From D. Kaplan, "The Development of Palm Leaves" *Scientific American* (July 1983): 98.)

Does the paragraph have a topic sentence? If so, underline it.

How is the paragraph developed?

If there is no topic sentence, write the main idea of the paragraph in the space below.

Can you suggest any way of making this paragraph clearer?

Activity 18.5

Why Do Yarrow Plants Vary?

Recall the yarrow plants graphed in Activity 18.1. The pictures and graphs there were taken from the following:

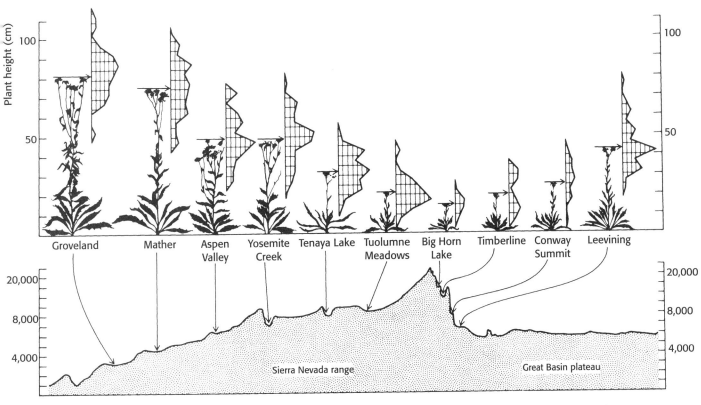

J. Clausen, D. D. Keck, and W. M. Hiesey, *Experimental Studies on the Nature of the Species. 111 Environmental Responses of Climatic Races of Achillea,* Carnegie Institution of Washington publication 581(1948).

Like the Groveland and Leevining populations in Activity 18.1, all these populations were grown under the same conditions as Stanford University (altitude 100 ft). Biologists grew seeds there in an effort to determine whether variation in plant populations was caused by genetic factors or by environmental factors. The lower graph shows the altitude and approximate geographic location from which each population originated. All the locations were at about 38° N latitude.

Based on this graph, do genetic factors or environmental factors seem more likely to determine plant height?

Examine the table below, which shows the growth and survival of plants from the same populations at three different locations. The altitude at Stanford is 100 ft, at Mather 4600 ft, and at Timberline 10 000 ft.

Growth and survival of ecotypes of A. *lanulosa* grown in experimental plots at Stanford (100 ft elevation), Mather (4600 ft), and Timberline (10 000 ft) in California for 3 years.

Origin of plants	Longest stems (cm)			Survival* (%)		
	Stanford	Mather	Timberline	Stanford	Mather	Timberline
Groveland	83.6	58.2	15.5	100	93	40
Mather	79.6	82.4	34.3	93	90	39
Aspen Valley	47.4	56.8	25.3	93	100	73
Yosemite Creek	42.6	56.2	30.1	97	97	90
Tenaya Lake	33.9	33.7	33.4	100	97	97
Tuolumne Meadows	24.5	32.7	28.4	90	97	93
Timberline	21.2	31.6	23.7	90	67	90
Big Horn Lake	15.4	19.5	23.6	83	67	91

*Based on samples of 30 plants (except Big Horn Lake, 12 plants). J. Clausen, D. D. Keck. and W M. Hiesey, *Experimental Studies on the Nature of Species. 111. Environmental Responses of Climatic Races of Achillea,* Carnegie Institution of Washington publication 581(1948).

What can you now conclude about genetic and environmental factors?

Write a paragraph describing the variation shown here as an adaptation to selective forces. What are those forces?

Activity 19.1

Identifying a Main Idea

  Chemists must be included among the garlic and onion lovers. For them the reasons are professional: chemists have long been attracted to substances that have strong odors, sharp tastes and marked physiological effects. Investigations made by chemists over more than a century establish that cutting an onion or a garlic bulb releases a number of low-molecular-weight organic molecules that incorporate sulfur atoms in bonding forms rarely encountered in nature. The molecules are highly reactive: they change spontaneously into other organic sulfur compounds, which take part in further transformations. Moreover, the molecules display a remarkable range of biological effects. The lacrimatory, or tear-inducing, quality of an onion is only one example. Certain extracts of garlic and onions are antibacterial and antifungal. Other extracts are antithrombotic, that is, they inhibit blood platelets from forming thrombi (aggregations of themselves and the protein fibrin). In short, they act to keep blood from clotting. (From E. Block, "The Chemistry of Garlic and Onions" *Scientific American* (March, 1985): 1 14.)

Read the paragraph above again and identify the main idea. Write the main idea in the space below.

Reread the paragraph. Do you agree that your main idea is the main idea of the paragraph? If you do not, rewrite your main idea.

Activity 19.2

Concept Mapping Photosynthesis

Reread the section on photosynthesis carefully and construct a concept map that illustrates the process of photosynthesis. Include as major subtopics the light reactions and Calvin cycle, and their end products, as well as the absorption of light energy, the conversion of light energy to chemical energy, and the storage of chemical energy as a sugar in the plant.

Activity 19.3

Photosynthesis vs. Cellular Respiration

One way to increase your understanding of a new process is to compare it with another, similar process that you already understand. By looking for similarities and differences you can recognize trends the processes may have in common. The trends in photosynthesis and cellular respiration illustrate some major theories that underlie all biology.

         Photosynthesis   **Cellular Respiration**

Similarities

Differences

Name three main biological theories that are illustrated by these two processes.

1.

2.

3.

Activity 19.4

Transplanting Photosynthesis

Chapter 19 contains many of the details of photosynthesis. Recall that many of the organisms in the various kingdoms cannot produce their own food by photosynthesis.

What important chemical compound are these organisms lacking? _____

What are the structures in which this photosynthetic compound is located? _____

Review the sections of Chapter 19 that deal with photosynthesis and make certain that you understand all of the raw materials that are needed to carry out this important process.

Genetic engineers have been working for several years trying to isolate genes that control photosynthesis. Imagine the genes have been located and the nucleotide sequences identified. Select a nonphotosynthetic organism in which to implant the engineered photosynthesis genes. In the space below, identify the organism you have selected and tell why you selected it. Explain the benefits that can be derived from enabling this organism to carry out photosynthesis. Explain how this organism will obtain all of the necessary raw materials to conduct photosynthesis in its cells. Try to counter any arguments you might encounter from persons who do not know anything about genetic engineering techniques.

Consider C_4 plants and the genes that could be available to genetic engineering. Which traits of a C_4 plant may you want to add to the genome of your organism?

Consider the unique adaptations of the CAM plants. What genes may you want to incorporate into your organism's genome?

What advantages would your final organism have to enhance its chances of survival?

What new selective pressures may it experience with its new characteristics?

What do you think would be its chance for survival?

Activity 19.5

Effect of Gibberellic Acid

A class of biology students conducted an experiment on the effect of gibberellic acid on the growth of two varieties of peas, Alaska and Little Marvel. The table below contains the measurements for five plants in each of four treatment groups. Study the data carefully.

Growth of Two Varieties of Pea Plants with and without Gibberellic Acid Treatment

Pea variety	Treatment	Individual plants	Length (mm) initial measurement	Length (in mm) on days following initial measurement				
				1st	2nd	3rd	4th	7th
Alaska	Sprayed with gibberellic acid (Experimental)	1						
		2						
		3						
		4						
		5						
		Average	67	101	130	153	179	214
	Sprayed with water (Control)	1						
		2						
		3						
		4						
		5						
		Average	61	84	105	121	139	163
Little Marvel	Sprayed with gibberellic acid (Experimental)	1						
		2						
		3						
		4						
		5						
		Average	35	51	76	93	103	156
	Sprayed with water (Control)	1						
		2						
		3						
		4						
		5						
		Average	32	38	43	51	56	79

From Addison E. Lee, 1968, *Plant Growth and Development* (Lexington, MA: D.C. Heath and Company). Copyright © BSCS.

What can be concluded from the results of the experiment?

The graph below was constructed by the students using the data from the table on the previous page. Observe the graph carefully.

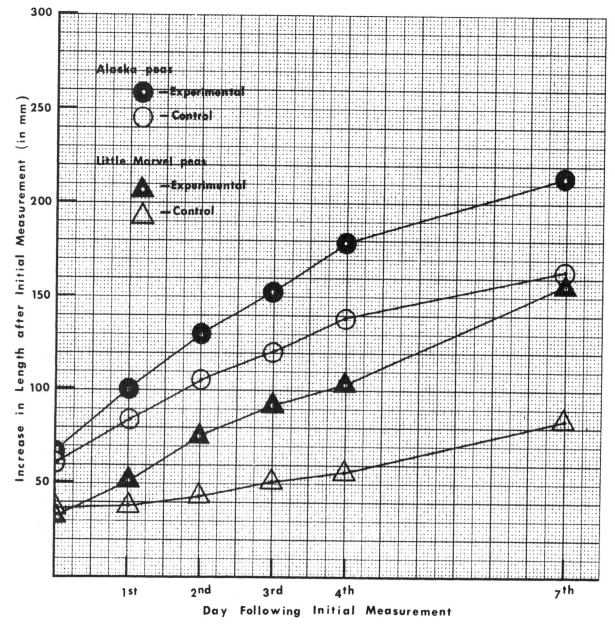

What can you conclude from the results of the experiment?

Which format is easiest to interpret—the table or the graph?

Explain why one is easier to interpret than the other.

Chapter 20

Activity 20.1

Looking Ahead

Perhaps you already know something about the ideas in this last section of the Green Version. In the space below, write continuously for five minutes about what you know, think you know, or would like to know, about these topics: behavior, natural selection, and survival; world ecosystems on land and in water; and human ecology.

Activity 20.2

Mimicry Hypothesis

Monarch butterflies lay their eggs on milkweed plants. When the larvae emerge, they feed almost exclusively on the milkweed plant. The milkweed manufactures several different alkaloid chemicals that produce a bitter taste. The monarch larvae consume large amounts of this chemical, and it is incorporated into their bodies. When the adult butterflies emerge, they still contain the bitter tasting chemical. Biologists have observed birds feeding on the monarch butterflies. Soon after eating one, the bird regurgitates violently and expels the butterfly from its mouth. Scientists assume that the bitter tasting chemical is the cause of the bird's regurgitation.

Biologists also have noticed that the viceroy butterfly closely resembles the monarch. One very popular hypothesis is that birds who have eaten monarch butterflies learn to avoid viceroy butterflies because they look like the monarch. The viceroy butterfly, however, does not contain the bitter tasting alkaloids the monarch does. The two butterflies are shown in the illustration below.

Design an experiment to determine the reason why birds avoid eating the viceroy. Is it because it mimics the appearance of the monarch, or is there some other bad-tasting molecule in the viceroy's body? In the space below, describe how you would conduct such an experiment. Identify the variables in the experiment and how you would control variables that might have an influence on the outcome of the experiment. What results would you expect to find that would support the mimicry hypothesis? What results would you expect to find that would not support the mimicry hypothesis?

Activity 20.3

Mating Rituals

Much of what scientists know about behavior has been gathered through observation, preferably in the organism's natural environment. The examples of behavior discussed in Sections 20.2 through 20.6 are descriptions of behavior that scientists have observed and recorded. Many of these behaviors can be observed in a variety of organisms.

Review the important role of sexual selection in many populations. What are the key principles of sexual selection?

Often we do not think of ourselves as animals with behaviors similar to other species. However, we do exhibit many of the same basic behaviors during courtship. You probably have observed this courtship behavior in your school halls before and after school, as well as between classes. Record some observations you have made of the courtship behavior of adolescent male and female *Homo sapiens* in their daily environment. Include descriptions of plumage or other attention attracting elements, aggressive and submissive behavior of males or females, and the sequence of steps or small behaviors that appear part of the ritual. Describe any characteristics that appear to be favored in the sexual selection process by the male and the female.

Did you see any examples of territoriality or cooperative behavior associated with these courtship rituals?

Activity 20.4

A Fish Model

Striped bass normally live in ocean waters and estuaries. They also can live in fresh water; in fact, they are used to stock rivers and ponds.

Biologists conduct many studies on striped bass, in part because these fish are prized as human food. The fish are tagged with temperature-sensing transmitters that send information about a fish's location and the temperature of the surrounding water. Such studies show that striped bass of different ages prefer different ranges of environmental temperatures. Juveniles prefer to be in water between 22° and 27° C; adults, between 16° and 25° C. This preference is so strong that a starving bass will not follow prey into water that is too warm or too cool. As the drawing shows, this causes a coastal population of fish to disperse.

Males reach reproductive maturity at 2 years, and females at about 4 years. Adults return to shallow water to spawn.

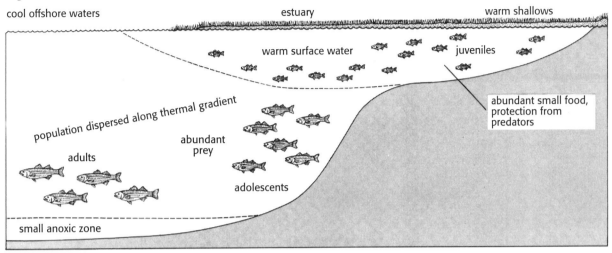

cool offshore waters   estuary   warm shallows

warm surface water   juveniles

population dispersed along thermal gradient

abundant small food,
protection from
predators

abundant
prey

adults

adolescents

small anoxic zone

The season also affects the water temperature. Eggs are laid and hatch in the spring, when water is cooler. The hatchlings reach the juvenile stage by summer, when the water is warmer. Meanwhile, the adults move out to deeper, cooler waters. Space and food resources thus are available to all members of the population.

The bass also are limited to areas that have desirable concentrations of oxygen. They must have a minimum of 2 mL of oxygen per liter of water. The drawing shows that in the deepest waters, too little oxygen is available. In inland ponds, the anoxic (low oxygen) area extends higher in the summer because of the greater decomposition of organic material in warm weather, which uses up oxygen.

Develop and use a model of growth for a striped bass population, similar to the models in the textbook investigations. Include assumptions about initial population size, depth of the oxygenated zone, geographic area (including coastal or inland), temperatures at various depths, time of year, and changes of temperature from seasonal changes. Decide whether the fish are free to migrate to warmer or cooler areas, or if they die when the water is too warm or cool. Whatever your model is like, the resulting output should be the numbers of adults and juveniles in a specified area at a specified time of the year.

Write up your model clearly so that your classmates or another person can use it without asking you any questions.

Chapter 21

Activity 21.1

In Search of Early Artists

On the walls and ceilings of a cave near Lascaux, France, are magnificent paintings. Some are far underground. In a great chamber, called the Hall of Bulls, the paintings are of gigantic bulls. In adjoining passages are paintings of horses, deer, wild cattle, and bison. Paintings in other parts of the cave show a man, a rhinoceros, cats, and other animals.

To determine who the artists were and when they lived, scientists studied the things the artists left behind. Stone palettes and lumps of pigments were some evidence they left, as well as the remains of meals. Because they had to paint in the dark, they used lamps made of stone. Tallow probably was the fuel, and the wicks were made of lichen or twigs of juniper.

Ashes and soot found with the lamps have provided carbon for radiocarbon dating. As any organism takes in carbon during its lifetime, it takes in a certain proportion of carbon-14 (^{14}C), a radioactive element that is formed in the air. When the organism dies, the ^{14}C begins to decay to nitrogen at a definite rate: about every 5730 years, half of the ^{14}C originally present has been converted to nitrogen. (In other words, the half-life of ^{14}C is 5730 years.)

By measuring the amount of ^{14}C in a fossil and comparing it to the amount that would be expected (based on the amount of ordinary carbon that is present), a scientist can determine the fossil's age.

Draw a graph showing the breakdown of ^{14}C that occurs over the years.

After how many years is less than 1% of the original ^{14}C left in a fossil? _____

The charcoal found with the lamps contained about 13% of the ^{14}C that was in the original fuel. How old are

the paintings? _____

Activity 21.2

Developing Hypotheses:
The Role of Plate Tectonics in Evolution

The crust of the earth consists of large plates that float on magma. As they float, the plates separate and collide with one another. The separation of plates results in the formation of new crust, called rifts, usually on the ocean floor. The collision of plates folds crustal material upwards, thus creating new mountain ranges. Mountain ranges, in the form of volcanos, also are formed where the crust is relatively thin. The continents move with the plates, adding and losing landmass as they separate and collide. The following figures depict the break-up of the supercontinent Pangaea that existed 225 million years ago.

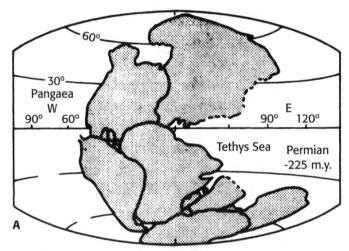

The ancient landmass Pangaea, meaning "all lands," may have looked like this about 225 million years ago, at the beginning of the Mesozoic. Pangaea was composed of two main sections: Laurasia in the northern hemisphere and Gondwana in the southern hemisphere. Panthelassa, meaning "all seas," evolved into the Pacific Ocean. The present Mediterranean Sea is a remnant of the Tethys Sea.

1. What continents composed Pangaea at the beginning of the Mesozoic Era?

Form a hypothesis that explains the relationship between the position of the continents and the distribution of plant and animal populations.

Locate North America on the map. Given its position at this time, what type of environment might you expect?

Notice that all the landmass is concentrated in one region of the earth's crust and the remaining surface is covered by seas. What role might this portion of the earth's history have played in evolution?

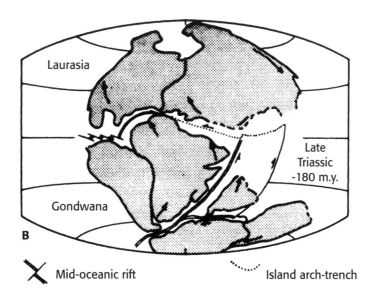

By the end of the Triassic, about 180 million years ago, continental drift changed world geography. New ocean floor was formed by spreading zones between some of the continents. Arrows depict the motion of the drifting continents.

2. By the end of the Triassic Period, which continents had separated from Pangaea?

What effect might this separation of continents have had on the evolution of plants and animals?

How has the climate of North America changed after the 20 million years of drift?

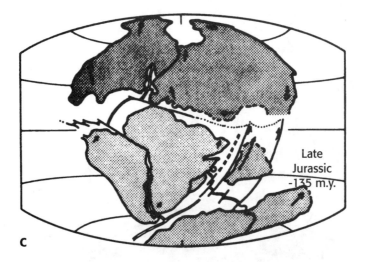

c

By the end of the Jurassic Period, 135 million years ago, world geography had changed even more. More ocean floor was created between North and South America, and between India and Gondwana.

3. By the end of the Jurassic Period, the continents had separated further. Which continents became isolated and how might this isolation affect the gene pool of populations?

What happened to the environment and climate of North America during the last 45 million years of drift?

What animals were present during this time and how might the environmental changes have affected them?

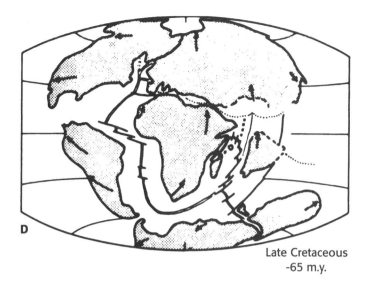

Late Cretaceous
-65 m.y.

By the end of the Cretaceous Period, 65 million years ago, most of the modern continents had separated. Large amounts of ocean floor were created in the Atlantic Ocean region.

4. Which continents could have exchanged populations, thus affecting the gene pools?

Which continents became isolated during the last 75 million years and how might this isolation have affected the evolution of populations?

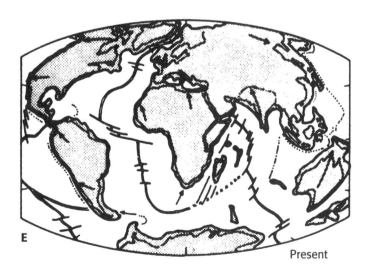

Present

The present world geography brought some of the continents together that previously were separated.

5. During the past 65 million years, which continents became isolated?

Are there some unique species as a result of this isolation? Give examples.

Which continents, previously separated, are reconnected by landmass?

What are some differences and similarities between some of the organisms now found on the continents?

Activity 21.3

Animal Families through Time

Much science data is summarized in tables, which saves space and makes it easier to do further calculations based on the data.

Like graphs, data tables usually are arranged along vertical and horizontal axes. At the top of a data table is a labeled horizontal row of one or more variables. At the side is a labeled vertical column of one or more variables. For example, the simple data table below has one variable in the horizontal row, and two in the vertical column:

Gender	Number of students
Boys	16
Girls	18

By reading the labels and numbers, you learn that 16 students are boys, and 18 are girls.

Most tables are more complicated and need more interpretation. In the following table, the results of four experiments are summarized. For each experiment, four results (A–D) are recorded. Notice that there also is a horizontal row for the average of each result for all four experiments.

Experiment	Results			
	A	**B**	**C**	**D**
1	4	5	20	0.9
2	2	6	25	0.4
3	3	5	27	0.5
4	2	7	24	0.6
Average	2.75	5.75	24	0.6

Construct a data table showing the number of animal families that have lived in past eras. Base the table on the information below, but use the names of eras and periods instead of (or in addition to) number of years. Also, show the types of animals that lived in each period.

Number of families	Million years ago	Number of families	Million years ago
850	0.011	300	310
810	5	310	320
620	63	320	345
510	106	320	365
360	135	290	390
320	148	280	405
310	157	270	411
280	180	250	418
270	187	230	425
240	192	210	431
300	200	180	487
310	256	100	500
300	280	80	533
300	290	60	566
295	300		

You also could show the information as a graph. What are the advantages of using a data table over using a graph?

What are the advantages of using a graph over using a data table?

Activity 21.4

Surviving the Cold

The Antarctic is too cold today for most fish; in the Southern Ocean, the temperature rarely rises above 2° C. However, a group of perchlike fishes called notothenioids live there.

Antarctica once was part of Gondwanaland. About 80 million years ago, that landmass began to break apart. For the next 40 million years or so, the area that now is Antarctica probably was surrounded with fairly warm water. Fossils of sharks, rays, catfish, and other fish now found only in temperate waters have been found on Seymour Island. When Antarctica separated from Australia, ocean currents caused enormous cooling of the new Southern Ocean.

The notothenioids have two adaptations that help them survive cold water temperatures. One adaptation is the ability to make substances that act as antifreeze in their blood. Because the notothenioids' kidneys lack glomeruli, the substances are not lost in the urine, but stay in the blood. Related fish that live in warm water have kidneys with glomeruli.

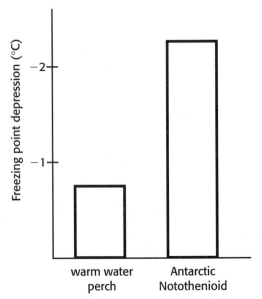

The histogram on the preceding page compares the antifreeze ability, measured in terms of freezing-point depression, of notothenioids and warm-water perches.

Most notothenioids are bottom-dwellers, but at least two species have another helpful adaptation, buoyancy that lifts them to middle levels, where water is warmer than at deep levels. There buoyancy is surprising, for, like their bottom-dwelling relatives, these notothenioids have no swim bladders. Their skeletons, however, contain a high proportion of cartilage, which is lighter than bone tissue. In addition, their vertebrae are hollow.

Write a paragraph about what you think happened to Antarctic fish about 30 million years ago.

Chapter 22

Activity 22.1

Solar Radiation

Each day the earth rotates once on its axis, and, depending on the season, most or all of the earth goes through a cycle of daylight and darkness. These photos taken from satellites show the western hemisphere at various times of the day.

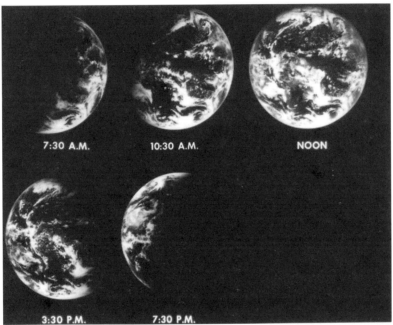

NASA

Use the photos as a reference to construct a line graph showing the approximate percent of the western hemisphere receiving solar energy at midnight, 7:30 A.M., 10:30 A.M., noon, 3:30 P.M., and 7:30 P.M.

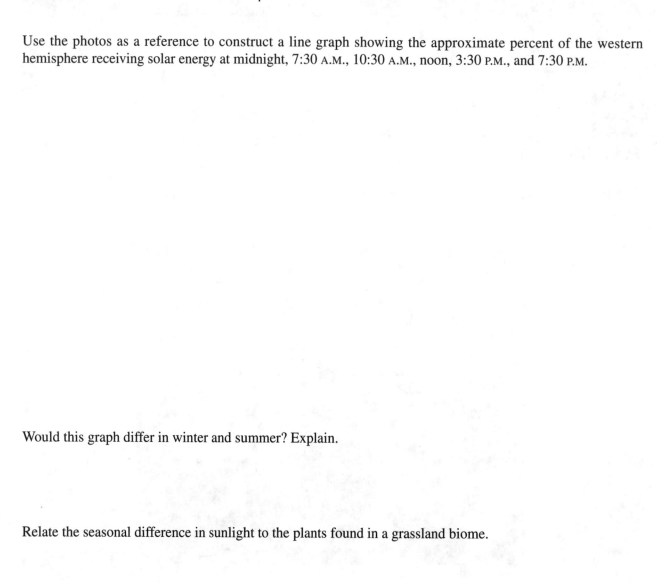

Would this graph differ in winter and summer? Explain.

Relate the seasonal difference in sunlight to the plants found in a grassland biome.

Activity 22.2

Reading About Biomes

You may find the SQ3R method helpful in mastering this chapter.
Follow these steps:

1. S: *Skim* the passage quickly.
2. Q: Note the major *questions* that you think are answered by the passage. Watch for clues in the printing, such as headings and boldface type. The authors used these to emphasize major ideas or vocabulary.
3. R: *Read* the passage carefully.
4. R: *Recite* the answers to the questions.
5. R: *Review* the passage. Find any answers you missed earlier.

For example, one student wrote these questions about Section 22.1:

1. What is climate?
2. Why is it important to ecosystems?
3. How does climate change in different seasons?
4. What is a climatogram?
5. What is a biome?

Try the method yourself on one of these sections: 22.3, 22.4, 22.5, or 22.8. After steps 2 and 4, fill in the spaces below.

Questions	Answers

Activity 22.3

Biomes in Three Dimensions

The solar radiation and precipitation an area receives depends on where it is located on the globe—on its latitude, its longitude, and its altitude. The drawing below shows a mythical continent that has characteristics of Eurasia and Africa. Mt. Echo, at 25 degrees of longitude and 40 degrees of latitude, is 5000 m high. For each 1000 m above sea level, the vegetation on the mountain changes as if it were 10 degrees of latitude farther north. For example, at 1000 m of elevation, the vegetation changes from grassland to deciduous forest.

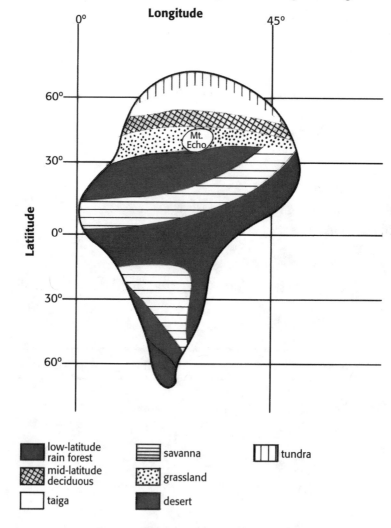

Use your textbook and map to answer the following questions. List at least five organisms in answer to each question.

What organisms would you find at sea level, 10 degrees north and 0 degrees east?

What organisms would you find at sea level, 0 degrees north and 25 degrees east?

What organisms would you find at 3000 m of elevation, 40 degrees north and 25 degrees east?

Activity 22.4

Where Biomes Meet

Three biomes overlap in central Wisconsin. At that latitude and longitude, climatic factors are such that the vegetation on the sunny side of a hill may be different from the vegetation on the shady side. The difference in temperature or precipitation can tip the balance from one biome to another.

Some areas can be described as follows:

1. In summer, herbs and grasses grow on the gently rolling hills. Bur oak trees are seen here and there, but they grow mainly in valleys, along streams. Other trees are rare. Rabbits and grasshoppers are common. The winters are cold, with an average temperature of –6° C in January; in July, the average temperature is 21° C. The greatest amount of precipitation falls from April through June.

What biome is described here? _____

2. In other areas, the greatest amount of precipitation falls between June and September. The temperature also is highest then, the average ranging from 10 to 15° C. The winters are bitter cold, with the average temperature well below –18° C in January. Most trees are adapted to conserve water—their leaves are needles. Evergreen forests of spruce, fir, pine, and hemlock surround lakes carved out by glaciers. Some deciduous trees also may be seen here, including beech and aspen. Deer and black bears live here.

What biome is described here? _____

3. Describe Wisconsin's third biome. Write a paragraph similar to those above. Name the biome and include information about the climate, plants, and animals. Use your textbook and other references.

Activity 22.5

Human Intervention

Natural succession in biomes is a slow process and decades may pass before there is any noticeable change. The first stages of succession in an environment that has been drastically changed, however, may quickly appear. For example, weeds and other opportunistic plants quickly establish themselves in vacant lots and along roadsides. Larger plants such as trees and shrubs usually take a longer time to become established. Humans often speed up the process by planting vegetation that eventually might become naturally established in the area. Is this a good idea?

Three examples of areas that have had some type of disturbance in the biome and the attempts humans have made to intervene are discussed below. Use the library to research the cause and extent of the disturbance, and what has been done to alter natural succession. Using this information, debate the policy of human intervention and decide a proper course of action.

Ethiopia

Research how desertification in this area occurs and whether it is part of natural succession or the result of human intervention. How have humans attempted to halt the desertification? Have these attempts worked?

Mt. St. Helens

Although the disturbance in this area is the result of natural causes, has the recovery been sped up through human intervention or is it natural? What natural succession has taken place?

1988 Yellowstone National Park Fires

What caused the fires? What was the policy regarding forest fires? In what ways did humans intervene? What natural succession has taken place since the fires? What have humans done to speed up the recovery of the forests?

Chapter 23

Activity 23.1

A Muddy Sea around Us

Science writing must obey the rules of spelling, punctuation, and grammar to be clear and communicative. Some mistakes may actually change the meaning of a sentence. For example, consider these sentences:

> Sewage greatly effects nearby ecosystems.

The word *effect* means "bring about." The author meant to say *affect*.

> Being polluted, the biologist studied the fish in the stream.

Presumably the stream, not the biologist, was polluted.

> The hydrosphere absorbs stores, and circulates heat.

This is a case of a missing comma. The hydrosphere does not "absorb stores"; it absorbs, stores, and circulates heat. In science writing, each sentence is similar to a chemical or mathematical equation its meaning must be perfectly clear to the reader. The selection below is adapted from the book *The Sea Around Us,* by Rachel Carson. Try to find and correct the errors introduced into the selection.

> Between the sunlight surface, waters of the open sea and the hidden hills and valleys of the ocean floor lie the least well-known region of the sea. these deep, dark waters, with all their mysteries, and there unsolved problems. Cover a vary considerable part of the Earth. The hole world ocean extends over about three-fourths of the surface of the globe. If we subtract the shallow areas of the continental shelves and the scattered banks and shoals, where at least the pail ghost of sunlight moves over the underlying bottom, there still remain about half the earth that is covered by miles-deep, lightless water, that has been dark since the world began.

Activity 23.2

Aquatic Ecosystems

What makes up an aquatic ecosystem? How are these ecosystems classified and what makes them different from one another? As you read or review Sections 23.1 through 23.9, construct a concept map that illustrates the classification and characteristics of the various aquatic ecosystems.

Activity 23.3

Aquatic Variables

Many newspaper articles contain descriptions of surveys or experiments in which variables are measured. Read the article below, released by United Press International in 1985, and look for variables.

State, MWD Widen Water Pollution Tests

SACRAMENTO (UPI) —The Department of Water Resources is widening the scope of its tests of Northern California water shipped to Southern California via the State Water Project to check for pollutants.

The Metropolitan Water District of Southern California, which serves most of the Southland's urban area, is also increasing scrutiny of its water, most of which comes from the Colorado River. Los Angeles uses some MWD water, but relies mostly on its own supply.

A water quality engineer for the Department of Water Resources said Friday that sediments in storage reservoirs will be checked for "pesticides, other organic materials and minerals, including selenium."

He said tissue samples will be taken from fish in the California Aqueduct and in the department's reservoirs in Southern California. "We want an assurance that things in fish tissue are not approaching official-action levels," Mitchell said.

The crisis over the selenium-tainted Kesterson Wildlife Refuge in Merced County has raised concern over the Northern California water supplied by the State Water Project.

Part of the water comes from the San Joaquin River, which flows into the Sacramento-San Joaquin River Delta. A director of the San Francisco Bay Institute said Friday that farm waste water from about 77,000 acres of land on the western side of the San Joaquin Valley drains into the river. He said selenium probably had been entering the San Joaquin for many years before the Kesterson crisis arose.

The MWD gets a relatively small portion of its water from this source.

In Los Angeles, an MWD spokeswoman said Friday that selenium content of the Colorado River water also is a matter of concern.

She said tests of MWD water supplied from the Colorado in January showed selenium contents of 4 to 6 parts per billion, while tests of water coming in from the north at the same time showed no detectable selenium content.

She emphasized that tests for selenium of both water from the north and water from the Colorado have always been below the federal government's action level of 10 parts per billion. (Reprinted with permission of United Press International, copyright 1985.)

What agency has conducted a study that involves variables?

What is the independent variable?

How does the independent variable vary?

What is the dependent variable?

Draw a histogram below to show the results of the study, and indicate how the results compare with a government standard.

Activity 23.4

Chesapeake Bay

Chesapeake Bay has been an important source of seafood since the first colonists arrived there. The food web depended on submerged grasses, and included crabs, oysters, ducks, striped bass, and other organisms. Farms and large cities eventually grew along the bay.

The grasses now are gone from Chesapeake Bay, and some of the fish and other organisms also are disappearing. The water near Cambridge, which was clear 25 years ago, is filled with mud. Large mats of algae float in the water and cut off light to the water below. Decomposers of dead algae further contribute to the depletion of oxygen in the water. Human waste sometimes can be seen in the water. Amounts of nitrogen and phosphorus have greatly increased, especially in the parts of the bay near the Susquehanna, Potomac, and James rivers. The areas where oxygen depletion is greatest are shown on the map in black.

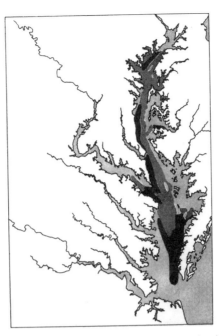

Reprinted with the permission of United Press International, Inc.

Write a paragraph that describes the probable sources of these changes in the bay, and how the sources made the changes.

Chapter 24

Activity 24.1

Cloudy Writing

The mean surface temperature of the earth gradually has risen, beginning in about 1750. During this same time, the amount of carbon dioxide in the atmosphere also has increased and scientists have tended to associate these two trends.

The warming trend has been spotty and is not evenly distributed over the earth. Computer simulations, however, have predicted an even warming trend in all areas. Scientists who developed some of the models commented on the contradiction as follows: "While we are witnessing a warming of the terrestrial climate, we cannot identify its cause. Even if it is of anthropogenic origin, it need not be due only to increased CO_2. But if it's due predominately to CO_2, then our present climate models require work to reconcile them with observational data on patterns of surface temperature change."

Rewrite this statement for clarity and correct any mistakes in spelling you find.

Activity 24.2

Reconstructing Village Life

Discoveries in archaeology and anthropology have led to a fairly good picture of what cities and villages in the Middle East were like about 4000 years ago. Cultural advances such as writing and metallurgy (metalworking with alloys or heated metals) made it possible to make many types of tools and to trade goods over long distances. Workers, such as butchers and potters, became specialized in their tasks. Artists created paintings and sculptures, and people decorated themselves with jewelry. There were several social classes based on wealth. Walls built around cities were for defense during times of war and also helped protect assets.

In contrast, about 8000 to 10 000 years ago, about the time of the Agricultural Revolution, early farmers probably lived very simply. Trading most likely took place within a small community, and tools were made with flint or shaped from cold metal. Buildings were constructed of stone, bone, and other unworked materials.

What happened during the time of transition, from about 8000 to 4000 B.C.? Our interpretation of this period largely is based on a village site in Iraq called Jarmo, excavated during the 1940s and 1950s. Jarmo existed about 7000 B.C. The people lived a simple life, farming the land around them. Each family was fairly independent and lived in its own house, made its own tools, and raised its own livestock. Archaeologists have found no evidence of the specialization of workers in Jarmo, nor have they found evidence of social classes or trade with outsiders.

After 1960, however, other sites were found that date from 6000 to 8000 B.C. These sites indicate that Jarmo was not typical. Some evidence from sites in Turkey, Jordan, and Iraq are presented below. What can you infer from this evidence? Write your inferences below each piece of evidence.

Wall painting of a hunter carrying a bow.

Large storage area associated with animals' bones, hunting equipment, and scraping tools; These remains are associated with large-scale leather-tanning.

Buildings that contain different types of craft tools—one with a butcher's tools, another a potter's, and so on.

Awls, pins, reamers, hooks, and sheets of copper that have been heated and shaped.

Large, sophisticated pottery kilns that could have provided many more pots than the local area could have used.

Obsidian (a volcanic glass) found hundreds of miles from any volcano.

Cities surrounded by walls and fortifications.

Differences among burial chambers. Some appear to be those of wealthier persons, some of poorer persons.

In a village in Turkey, a huge blood-stained stone slab surrounded with hundreds of human skulls.

Cosmetics in homes. Also, murals and figurines that depict people with elaborate hairdos and jewelry.

Sculpture of a head found in Jericho. The eyes of the head are made with shells that probably came from the southern coast of Israel, at least 100 km away.

Lumps of iron that had been heated.

Activity 24.3

Prediction–Extinction?

Cheetahs, the fastest animals in the world, appear to be headed for extinction. They once were found throughout the world, but now they live only in the parts of Africa shown in the map below.

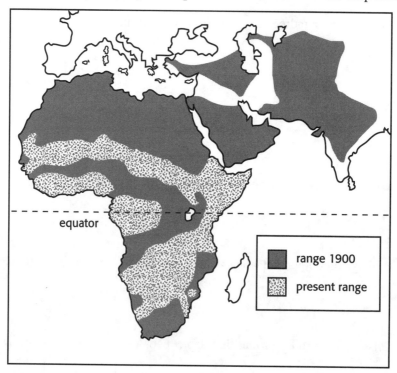

Several factors have led to this reduced population size. Apparently at some time in the Pleistocene, the population size was reduced severely and the inbreeding that followed reduced the genetic variation in the population. All living cheetahs are virtually identical genetically. Another factor in the decline of the cheetah population is environmental change.

List four dramatic changes that have occurred in Africa in the past 100 years that might cause a decrease in the range of the cheetah.

Previous chapters discuss the role genetic variation plays in an organism's ability to adapt to changes in the environment. Write several paragraphs about what might happen to the remaining cheetahs during the next 50 years. Consider the possible effects of genetic uniformity, climatic variations, changing ecosystems, and disease. What efforts could be made through protective legislation, research, and long range environmental planning that might influence the outcome of your prediction?

Activity 24.4

Predicting Your Future—Part II

Look at the predictions you made in the introductory activity. After studying biology this year, would you change any of these predictions? List any predictions you would change and explain your reasons for changing them. Also list some things you can do to help ensure the type of future you wish to have.

Arguing Your Ideas

Some scholarly articles appear to be mostly factual. Underlying most of these papers, however, is a thesis. A thesis is the writer's main statement, which he or she defends by producing evidence to support it. Usually the central thesis of a paper can be stated in one or two sentences. For example, Darwin presented his thesis—that evolution has occurred as a result of natural selection—in his book *The Origin of Species by Natural Selection.* The book contains the evidence he gathered and examined during a 20-year process of thinking and writing. Most theses are proposed and defended in considerably less time, but few are so thoroughly supported.

Try writing a biological thesis of your own. Use the entire textbook, your lab and field work, and any other sources you wish as the sources of evidence to support the thesis.

Some examples of theses proposed by students are the following:

If the organisms in an area are removed, the same organisms will eventually return.

Competition is usually beneficial to both competitors.

Or, take someone else's thesis (from discussion in the textbook or from a magazine or newspaper article) and present evidence that falsifies it.

Examples of theses you might argue against are:

The "population problem" is entirely a problem of distributing resources fairly.

Intelligence is determined entirely by genes.

Intelligence is determined entirely by environmental factors.

Your paper should be at least two pages long. The activities you completed earlier will help you in doing the research, writing your outline and rough draft, and finishing the paper.